JN314255

「続」英語で学ぶ生物学

── 生物科学の新しい挑戦 ──

The Second Volume of
Endeavors in Biological Sciences
── New Challenges in Biological Sciences ──

Ph. D. 渡邉　和男 編著
渡邉　純子 著

コ ロ ナ 社

Preface

科学技術の進歩は目覚ましく，日進月歩を超えた展開が日々生まれている。一方で基礎的な知見の確認と研究の忍耐強い積み重ねにより，科学技術の進展があることも忘れてはならない。生物学においては，学問の基本を理解するとともに，新しい知見の理解推進は必須のものである。今日では生物を見ることから生物から得られる多様な情報を利用するように変ってきており，日々大量の情報を検索し，研究に臨む必要がある。

先端的な研究課題については，日本国内の研究においても専門論文やインターネットでの情報は，世界共通言語ともいえる英語での記述が主流である。世界の権威ある科学雑誌や研究機関は当然ながら英語で情報を公開しており，英語での学問の理解は必須である。

本書は，前作の「英語で学ぶ生物学」を受け継ぎつつ，最近のトピックでありまた基礎的な事項としても理解が必要な題材を新たに提供している。それは幅広く題材を理解するための基礎的な情報を提供するとともに，また専門用語に慣れることを意図している。さらに生物学の知識の修得だけではなく，生物学に関わる倫理や社会の問題を同時に考察することも考慮している。

取り上げた各題材は，著者が世界の著名な研究雑誌や最近のニュースとなった話題から選抜したものであり，これらについて関連論文や書籍を精査し，その内容を独自にまとめたオリジナルなものである。世界中の多様な話題をすべて網羅するのではなく，日本と世界に共通した事項を選抜している。

各章はオムニバス形式でそれぞれ独立しているため，題材はどこからでも拾い読みすることができる。興味のある方々には，本文の後に掲載された参考書籍や文献を読むことで，さらに理解を深めていただければと期待している。なお基礎的な知見や専門用語については，「英語で学ぶ生物学」や，「英語で学ぶ医科学入門」「英語で学ぶ環境科学」（英語で学ぶ自然科学シリーズ）などを並行して利用いただければ，さらに理解が進むことだろう。

科学技術は将来への多様な可能性を生み出すが，一方では，得られた知見の正しい適用や社会との透明性のある合意形成がなければ，人類が適正に享受することはできない。これは生物学においてもしかりであり，本書で取り上げた各章のExercisesでは，単純に知見を英語で理解するだけではなく，知見の社会との関わりについて話題を提供し，学習グループや個別の読者がそれについて十分に考えることを期待している。

最後に，本書執筆にあたり長い期間ご支援くださった多くの関係者各位に感謝いたします。そして，本書が生物学およびその新しい知見利用の理解推進に資することができればと期待します。

2013年9月

著　者

Contents

Chapter 1
Discovery of Induced Pluripotent Stem Cells (iPS Cells)

Many of you have probably learned about planarians in a biology class. A planarian is an aquatic organism about 1 cm long that lives in cold water. Interestingly as in **Figure 1**, if you cut a planarian into four equal parts, the parts will regenerate to produce four planarians! In other organisms too, such as lizards or newts, regeneration occurs at areas where they have been cut. Even in humans, if you scratch your hand, the bleeding eventually stops and the scratched area slowly regenerates back to its original form. Among cells, the birth of these new cells results from the action of stem cells.

Stem cells are present in a variety of organisms. In a planarian, stem cells aggregate at wound openings and repair begins, after which a planarian completely identical to the original one is formed. Stem cells exist in humans as well, serving to replace damaged or aged cells or to newly replenish cells lost as a result of disease or injury. The only difference between stem cells in humans and those in planarians is that although human stem cells can regenerate, they are limited to producing only certain types of cells. Although a simple hand skin injury can be repaired, the restoration of large and complex parts such as whole arms and legs is impossible. These cells are known as adult stem cells.

In contrast to these adult stem cells, artificially created stem cells have also emerged on the scene. These cells, unlike adult stem cells, can turn into all manner of cell types. In 2006, a cohort of Japanese researchers generated the world's first iPS cells (induced pluripotent stem cells) from mice. These iPS cells, which were

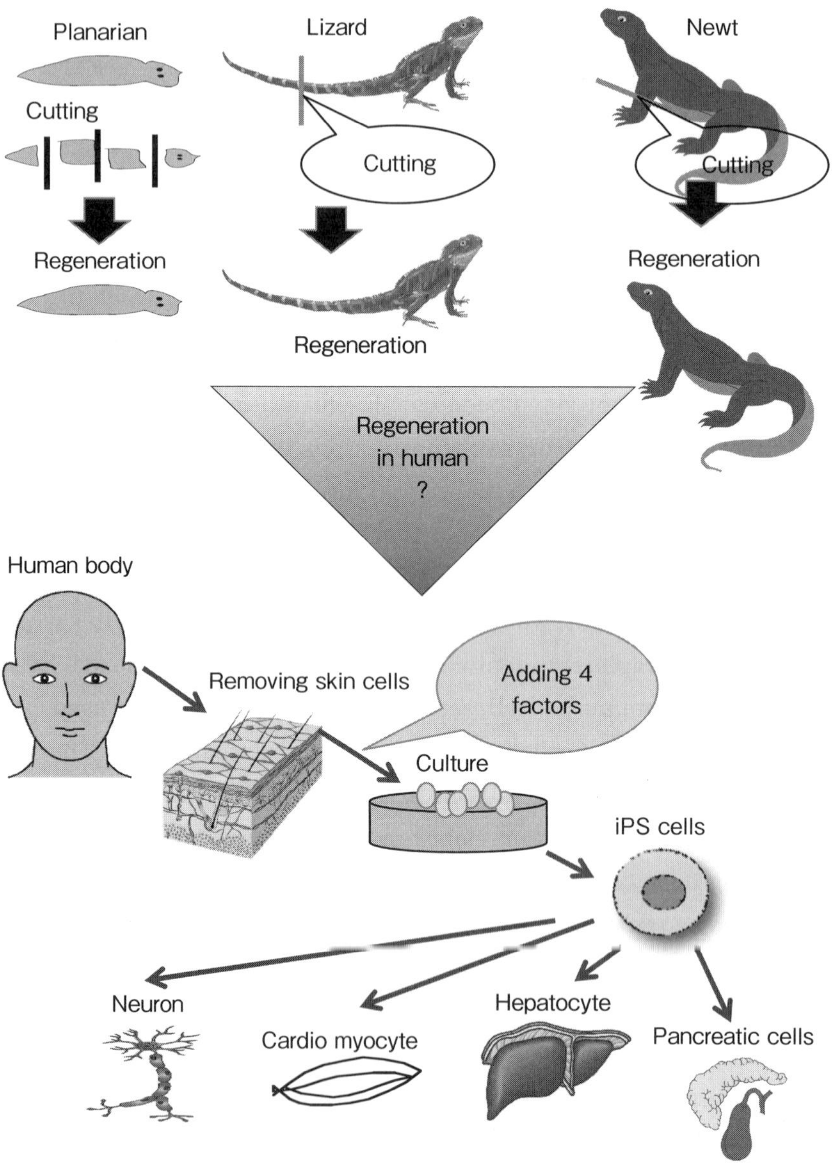

Figure 1 An idea for creating iPS cell production.

cultured from skin cells (fibroblasts) in mice, differentiated into cardiac muscle cells that beat in cell clusters. Two years later, the same technique was applied to cells obtained from human skin. News of this raced around the world in a flash.

These iPS stem cells differed greatly from the embryonic stem cells (ES cells) that had been used up until that time. Because fertilized eggs are not used to create iPS cells, the ethical barriers to their use are very low. In addition, immunological rejection is assumed not to occur because an individual's (the patient's own) skin cells are used.

A major catalyst for the discovery of iPS cells was the birth of Dolly the sheep in 1996. The nucleus taken from a mature udder cell of a sheep was inserted into an egg whose nuclei had been removed. When this egg was placed into the uterus of a different sheep, the newborn was completely identical to Dolly's genetic mother. This mysterious art produces what is known as a cloned embryo. This technique spectacularly overturned the conventional view until then that in mammals, cells that had differentiated would never rewind (become initialized). Four years later, in 2000, ES cells were produced.

A lizard's tail regenerates beautifully even if it is cut off. A newt's arm regenerates not only skin tissue but also bone when it is severed. Furthermore, as for planarians, they regenerate even after being cut into small pieces. In contrast, in mammals, starting from the time the egg is fertilized, cells become increasingly differentiated, finally turning into different organs, and it was thought that going in reverse was impossible; that is, that cells, once differentiated, could not possibly return to the initial egg cell, and that the process was a one-way street (irreversible). Nonetheless, the birth of Dolly and the production of ES cells demonstrated that this process could be rewound. Thus, the path toward the birth of iPS cells had begun.

The method of producing iPS cells is surprisingly simple. It is said that anyone with experience in basic genetic manipulation techniques and culturing may be able to produce them. However, it is taking time for this method to become established. Particularly important aspects are the genes that need to be introduced into non-pluripotent cells to force them to become iPS cells. Studies on ES cells thus far provided numerous hints, and the to-be-introduced genes were ultimately narrowed down to four (the Yamanaka factors) : those that 1) build the body, 2) play a key role in expression, 3) play an important role during regeneration, and 4) control transcription. Each of these genes is introduced into the cell via a *retrovirus*. Genes for 3) can carry a risk of causing cancer, but these can now be avoided.

Although the method of using *retroviruses* to introduce genes has great advantages, some genes cause adverse effects on the body, particularly in gene therapy. Hence, methods to directly produce iPS cells without the insertion of genes are also being explored.

The most innovative aspect of iPS cells is that they can be used for transplants and regenerative medicine. Given that iPS cells are cultured from the patient's own skin, there is unlikely to be immunological rejection. Currently, ideas such as preparing pools of cells according to the type of *human leukocyte antigen* (HLA) are being considered so that they can be used immediately when needed. These iPS cells could be applied to the treatment of various diseases. In mice, researchers have succeeded in producing blood stem cells from iPS cells and applying them to the treatment of *sickle-cell anemia*. There are also reports that transplanting neural stem cells derived from iPS cells into mice artificially given spinal injuries

---

Notes：*retrovirus*「レトロウイルス」, *human leukocyte antigen* (HLA)「ヒト白血球抗原」 *sickle-cell anemia*「鎌型赤血球症」

reduced the symptoms by 20 %. iPS cells, which are capable of becoming various tissues, are being considered for practical use in the treatment of diseases such as *Parkinson's disease*, *muscular dystrophy*, *myocardial infarction*, and diabetes, and positive outcomes of such work have been confirmed. Research is being advanced at core universities in Japan, beginning with experimental animals such as mice and working toward eventual application in humans.

Conversely, iPS cells can also help reveal the causes of various diseases. By producing stem cells from the cells of diseased patients, it will become possible to study how and why diseases occur when they do. Until now, tissues have been directly taken from diseased patients and cultured; but such cells hardly ever proliferate, and this is also a great burden for the patients. However, by using iPS cells derived from the cells of diseased patients, we are about to understand how pathogenesis occurs. For example, the increasingly rapid cell loss that occurs in *amyotrophic lateral sclerosis* (ALS) has been revealed visually as well.

iPS cells are also expected to bring about ground-breaking progress in drug research. In the past, candidate drugs were first screened in experimental animals such as mice, monkeys, or pigs and their effects were then tested by experimentally administering them to humans. However, a drug that was highly effective in animal experiments might show a reduced effect, a rejection response, or toxicity when administered to people. An enormous amount of time and effort is considered necessary in order to discover just one drug in the conventional way; however, if iPS cells are used, animal experiments will become unnecessary. By transforming the cells into

---

Notes : *Parkinson's disease*「パーキンソン病」
muscular dystrophy「筋ジストロフィー」, *myocardial infarction*「心筋梗塞」
amyotrophic lateral sclerosis (ALS)「筋萎縮性側索硬化症」

the desired tissue, the effects and toxicity of a drug can be confirmed.

Looking ahead, the day we can culture iPS cells in three dimensions and replace organs that have weakened or stopped functioning is not so far off. And though efforts to create human organs in the bodies of pigs have begun, there are many more problems that yet remain. To this end, there are many remaining challenges to be addressed concerning research on stem cells (including iPS cells), such as the consideration of ethical issues, establishment of systems and guidelines, financial support, and the understanding of the patients and general public.

References

1. Books

1）山中伸弥，畑中正一：iPS 細胞ができた！　ひろがる人類の夢（集英社文庫）(2008)，ISBN978-4-08-781395-1

2）山中伸弥，中内啓光 編：再生医療へ進む最先端の幹細胞研究，実験医学増刊 **26**-5，羊土社（2008），ISBN978-4-758-10289-6

3）田中幹人：iPS 細胞　ヒトはどこまで再生できるか？，日本実業出版社（2008），ISBN978-4-534-04384-9

4）升井伸治：SUPER サイエンス　iPS 細胞が再生医療の扉を開く，シーアンドアール研究所（2009），ISBN978-4-86354-039-2

5）山中伸弥：iPS 細胞の産業的応用技術，シーエムシー（2009），ISBN 978-4-7813-0122-8

6）中辻憲夫ほか監，梅澤明弘ほか編：再生医療の最前線　2010-ES・iPS・組織幹細胞の特性の理解と分化誘導，創薬・臨床応用に向けた品質管理，安全の基盤技術（実験医学増刊 Vol. **28**-2），羊土社（2010），ISBN978-4-7581-0304-6

7）杉本正信，帯刀益夫：細胞寿命を乗り越える ES 細胞・iPS 細胞，その先へ，（岩波科学ライブラリー 164），(2009)，ISBN 978-4-00-029564-2

8）阿形清和，中内啓光，山中伸弥，岡野栄之，大和雅之：再生医療生物学（現代生物科学入門 第 7 巻），岩波書店（2009），ISBN978-4-00-006967-0

2. Journal

1）K. Takahashi, K. Tanabe, M. Ohnuki, M. Narita, T. Ichisaka, K. Tomoda and S.

Yamanaka：Induction of pluripotent stem cells from adult human fibroblasts by defined factors. Cell **131** (5), pp.861–872 (2007)

3. URL (2013年9月現在)

1) http://www.icems.kyoto-u.ac.jp/cira/j/index.html「山中伸弥，石井哲也 監　幹細胞ハンドブック―からだの再生を担う細胞たち　京都大学iPS細胞研究所「CiRA」(2009)」

Exercises

1．つぎの1～52の語に対応する英単語または英熟語を本文から選び出し，発音しなさい（動詞は原形を記入しなさい）。

	Japanese	English		Japanese	English
1	プラナリア		14	回復	
2	生産する		15	人工の	
3	生物		16	創成する	
4	トカゲ		17	繊維芽細胞	
5	イモリ		18	マウス	
6	ヒト		19	分化する	
7	削る，擦りむく		20	心筋	
8	出血（する）		21	塊	
9	幹細胞		22	胚	
10	集計する		23	授精する	
11	補充する		24	免疫拒絶	
12	疾病		25	触媒	
13	傷		26	核	

	Japanese	English		Japanese	English
27	成熟（した）		40	培養	
28	同一の		41	再生	
29	子宮		42	処理	
30	遺伝的		43	乳房細胞	
31	クローン		44	脊髄の	
32	転回		45	症状	
33	器官		46	糖尿	
34	巻き戻し		47	増殖する	
35	操作		48	病原性	
36	転写		49	医薬	
37	移植（する）		50	投与（する）	
38	神経の		51	拒絶反応	
39	挿入		52	毒性	

2. つぎの各文が本文の内容と一致するものにはT(True)，一致しないものにはF(False) を，文末の（　）に記入しなさい。

（1）If you cut a planarian into four equal parts, the parts will regenerate to produce four planarians.（　）

（2）Stem cells are not present in a variety of organisms.（　）

（3）The iPS stem cells are not different greatly from the embryonic stem cells.（　）

（4）The iPS cells are also expected to bring about ground-breaking progress in drug research.（　）

（5）There are not many remaining challenges to be addressed concerning research on stem cells.（　）

3. つぎの日本語の各文を（　　）の中の語を用いて英語の文にしなさい（必要があれば動詞を適切な形に変換しなさい）。

（1）日本人の科学者が，世界で初めてマウスから iPS 細胞を作成した。(iPS cells, generate)

（2）医学研究従事者は，患者の個人情報について倫理的に考慮する必要が常にある。(engage, ethical, patient)

（3）1996 年における羊のドリーの誕生は，iPS 細胞の体系的作成を有名にした。(systematic generation)

（4）iPS 細胞由来の神経幹細胞を移植することで，人工的に脊髄損傷を持ったマウスは症状が 20 % 軽減された。(transplant, reduce, injury)

（5）目的とする組織に iPS 細胞を変換することで，新薬候補の効果と毒性が検証できる。(transform, target)

4. つぎの各問いに英語で答えなさい。

（1）How do you think the discovery of the iPS cells?

(2) How do you generate the iPS cells?

(3) You cut a planarian into four equal parts, what would happen?

(4) Do you agree with the artificial creation of living organism?

(5) Discuss what would be the future of the iPS cell research. Who would be the beneficial of the research outcome.

Chapter 2
Extermination of Malaria

Malaria occurs when a person is bitten by an *Anopheles* mosquito that carries the malaria parasite (*Plasmodium falciparum*) (**Figure 2**). If a person becomes infected, symptoms such as fever, headache, and nausea occur; if severe, this disease can lead to death. The malaria parasite is present in the salivary glands of *Anopheles* mosquitoes, the vector of the disease. The female mosquito needs to feed on blood for her eggs to develop.

When a person is bitten, the malaria parasite enters the person's body along with the mosquito's saliva. Once the malaria parasite enters the bloodstream it is carried to the liver where it penetrates a hepatocyte and multiplies. When their number has increased to a few thousand, the parasites destroy the hepatocyte and are released into the bloodstream, after which they invade the red blood cells and multiply (by asexual reproduction).

As the cells repeatedly undergo asexual reproduction, the cells transform into gametocytes with male and female distinctions (ready for sexual reproduction). However, fertilization cannot occur within the human body; it can only occur within *Anopheles* mosquitoes that have ingested the parasites while feeding on this blood. In this way, the parasite reproduces by circulating through *Anopheles* mosquitoes and humans.

According to World Health Organization (WHO) in March, 2013, About 3.3 billion people — half of the world's population — are

---

Notes：*Anopheles*「ハマダラカの属名」, *Plasmodium falciparum*「ヒトマラリア原虫」
red blood cells「赤血球」

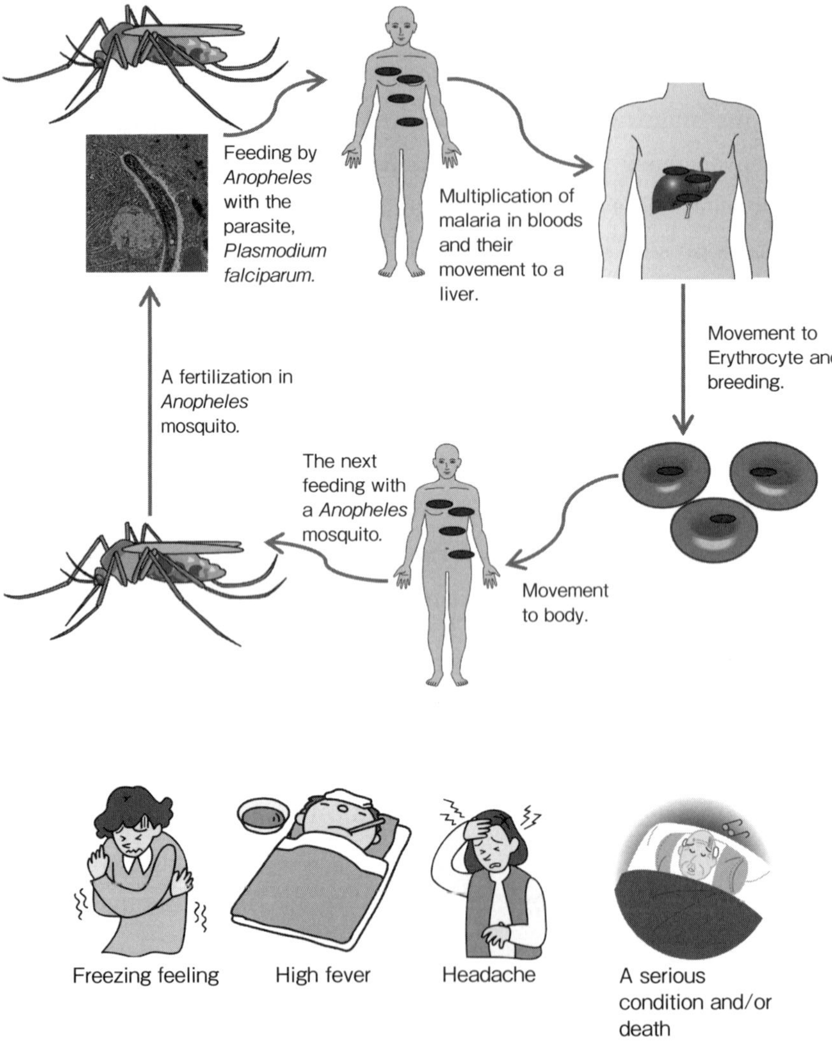

Figure 2 The lifecycle of malaria and its infectious symptoms in a human.

at risk of malaria. In 2010, there were about 219 million malaria cases (with an uncertainty range of 154 million to 289 million) and an estimated 660 000 malaria deaths (with an uncertainty range of 490 000 to 836 000). Increased prevention and control measures have led to a reduction in malaria mortality rates by more than 25 % globally since 2000 and by 33 % in the WHO African Region. The mortality rate is especially high for infants. The overwintering area of *Anopheles* mosquitoes is expanding as a result of global warming, and in Australia alone, the number of people infected in the year 2000 had increased to four times the number in 1970. Reducing deaths due to malaria is one of the long-held dreams of the human race.

One countermeasure for this purpose is the development of a malaria vaccine. Research into the development of a malaria vaccine has spanned several decades, but at long last has entered its final stages and a malaria vaccine is now at the stage of clinical trials in humans. One of the most promising vaccines at present is "FSV-1," which was originally developed by the United States Navy. It was first administered to adult men in Africa (in a study known as Malaria 025), and a clinical trial on an unprecedented scale with more than 2 000 children aged 1 to 5 is about to begin.

There are other methods to prevent malaria infection as well. One way is to avoid direct physical contact with *Anopheles* mosquitoes. To this end, the traditional mosquito net has been re-evaluated. A certain Japanese scientific manufacturer developed a special mosquito net. The company focused on two issues pertaining to the development of this net. They first considered the spacing between the threads. A mosquito will be able to get inside the net if this spacing is too large, whereas the ventilation will be poor if it is too small. A 4 mm net spacing was found to be just the right size to exclude *Anopheles* mosquitoes.

Next, they considered the material for the threads of the mosquito net. Mosquitoes will always land on the outside of a mosquito net, and with a normal mosquito net, a person stepping out has a high likelihood of being bitten and becoming infected. However, this newly developed mosquito net was woven from a thread that contains an insecticide. This insecticide, obtained by slightly modifying the active ingredient in the natural insect-repelling plant *Chrysanthemum cinerariaefolium*, is innocuous to humans. Although the net was originally developed for Japanese greenhouses, it did not gain wide acceptance in Japan.

Anopheles mosquitoes alighting on the mosquito net die as the drug gradually takes effect. Moreover, a great advantage of this mosquito net is that it continues to be effective even after it is washed. Another manufacturer had previously developed a mosquito net based on a similar concept; however, the insect repellent came off when the net was washed. So this new type of mosquito net was manufactured from specially processed thread. The insect-repelling effect of this mosquito net continues to last for at least two years, even with repeated washing.

After development of the mosquito net was complete, there were two problems in popularizing the net in developing nations. The first was the absence of any custom in the tropics of sleeping in a mosquito net. Another problem was whether the users would understand that the insect-repelling effect would continue to last even after washing. This mosquito net had been ordered originally by the Japanese government for aiding developing countries, but without subsidies it did not sell at all.

However, the situation drastically changed when the WHO

Note：*Chrysanthemum cinerariaefolium*「除虫菊」

recommended the efficacy of this mosquito net. In 1998, the WHO and UNICEF, among others, initiated the "Roll Back Malaria" campaign, setting the goal of halving the number of infections and deaths by 2010. The special threads used were sent out to various places around the world. Currently, they are being sent to areas south of the sub-Saharan region, primarily to places such as Ethiopia, Sudan, Uganda, and Tanzania. In some cases, the mosquito nets are locally manufactured and sold.

As yet another method, attempts to use biotechnology to prevent *Anopheles* mosquitoes from breeding have also begun. If *Anopheles* mosquitoes are rendered sterile by manipulating their genes, they will leave no progeny and the mosquitoes will decrease. This is also commonly debated at biodiversity conferences but careful risk assessment has been extensively applied to the transgenic *Anopheles* with sterility by many research organizations.

With the spread of malaria vaccines and mosquito nets, or with methods based on genetic manipulation, the day that malaria is eradicated may not be far off. The dawn of great new era is on the verge of breaking over various places around the world.

References

1. Books

1) 環境問題を考える編集者の会：レジ袋がなくなる日，マイクロマガジン社（2007），ISBN978-4-89637-250-2
2) 滝　順一：エコうまに乗れ！，小学館（2009），ISBN978-4-09-387852-4
3) ロバート．S. デソウィッツ：マラリア vs. 人間，晶文社（1996），ISBN978-4-7949-6254-6
4) 橋本雅一：世界史の中のマラリア―微生物学者の視点から―，藤原書店（1991），ISBN978-4-938-66121-2

Note：sub-Saharan region「サハラ砂漠以南の乾燥地」

5）岡田晴恵：感染症は世界史を動かす（ちくま新書）（2006），ISBN978-4-480-06286-4
6）マルティーヌ・モーレル：旅行者のためのマラリア・ハンドブック，凱風社（1998），ISBN978-4-773-62211-9
7）茂木幹義：マラリア・蚊・水田―病気を減らし生物多様性を守る開発を考える．海游舎（2006），ISBN978-4-905-93008-2

2. Journal

1）B. Maher：マラリアワクチン完成に向けて（The End of The Beginning.）pp.11-17．NATURE DIGEST（日本語版）（2008），ISBN1880-0556
2）M. Stopfer：Mosquitoes Bamboozled（News & views）. Nature 474, pp.40-41（2011）

3. URL（2013 年 9 月現在）

1）マラリア専門研究誌　http://www.malariajournal.com/content/
2）国立感染症研究所 2005。感染症の話 マラリア。
http://idsc.nih.go.jp/idwr/kansen/k05/k05_04/k05_04.html
3）環境省 2008。温暖化の感染症に係る影響。
http://www.env.go.jp/earth/ondanka/pamph_infection/full.pdf
www.env.go.jp/policy/kenkyu/suishin/backnumber/.../Kurane.pdf

Exercises

1．つぎの 1 ～ 40 の語に対応する英単語または英熟語を本文から選び出し，発音しなさい（動詞は原形を記入しなさい）。

	Japanese	English		Japanese	English
1	絶滅		6	唾液腺	
2	咬む		7	媒体	
3	寄生生物		8	血流	
4	感染（する）		9	浸透する	
5	頭痛		10	肝細胞	

	Japanese	English		Japanese	English
11	増殖する		26	殺虫剤	
12	無性の		27	成文	
13	繁殖		28	防虫	
14	有性の		29	無害の	
15	生殖母細胞		30	温室	
16	摂取する		31	降りる	
17	循環する		32	一般化する	
18	幼児		33	助成，補助金	
19	越冬		34	効能	
20	対策		35	与える	
21	周期，幅		36	操作する	
22	臨床試験		37	討論	
23	かってない		38	根絶する	
24	関係する		39	夜明け	
25	糸		40	寸前，間際	

2. つぎの各文が本文の内容と一致するものにはT(True)，一致しないものにはF(False)を，文末の（　　）に記入しなさい。

（1）Malaria occurs when a person is bitten by any mosquito species.（　　）

（2）Malaria causes problems only at developing countries.（　　）

（3）Sexual reproduction of Malaria only occur within *Anopheles* mosquitoes.（　　）

（4）Mosquito net does not protect at all from infection of malaria.

(　　)

(5) Transgenic *Anopheles* is available for testing reduction of the insect population. (　　)

3. つぎの日本語の各文を（　　）の中の語を用いて英語の文にしなさい（必要があれば動詞を適切な形に変換しなさい）。

(1) もしマラリアに感染すると，発熱，頭痛や吐き気などの症状が見られる。(nausea, symptom, infect)

(2) マラリア感染を防ぐ他の方法もある。(method, prevent)

(3) マラリアワクチン，蚊帳や遺伝子操作などにより，マラリアが根絶される日は，さほど遠くないだろう。(eradicate, manipulation)

(4) この薬は，多数の成人に投与され，臨床試験はかってない大規模なものである。(scale, unprecedented, clinical trial)

(5) WHO の支援による予防と管理措置は，世界的にマラリアによる致死率を 2000 年以来 25 ％減少させた。(prevention, reduction, mortality, globally)

4. つぎの各問いに英語で答えなさい。

（1）Where does the malaria reside within the vector insect?

（2）What is the innovative genetic property that transgenic mosquitos furnish?

（3）Would the mosquito net protect you fully from malaria?

（4）Is there any natural remedy to help prevention or cure from Malaria?

（5）Can on science discipline solve the malaria infestation?

Chapter 3
New Types of Influenza：Fear of Pandemics

It may be said that the history of mankind is a history of battles with infectious diseases. In ancient times there were the Plague (*Yersinia pestis*) and smallpox, and in modern times there have been periods during which tuberculosis, syphilis, and cholera were widespread. However, with the development of science, the range of infections gradually decreased. Furthermore, since the discovery of the antibiotic penicillin, our fear of tuberculosis, which was once considered incurable, has nearly disappeared. Other various antibiotics have subsequently been discovered as well, and the day we are liberated from the curse of bacterial infection appears to be getting closer. The battle against infectious diseases has seemingly come to a close.

However, in recent years, new fears are bearing down upon mankind. One is the battle against viruses far smaller than bacteria. A virus cannot survive on its own and must parasitize some organism. Influenza, in particular, is a virus intimately related to the lives of people and livestock (**Figure 3.1**). Influenzas periodically become widespread, but in developed countries, the mortality rate has greatly declined even if one becomes infected, owing in part to the efficacy of modern drugs.

Yet at this point, a novel influenza ("new-type influenza") epidemic is advancing on mankind. Particularly in the last few years, avian influenza (bird flu) has infected people and led many to their deaths. Furthermore, in the early summer of 2009, this new-type influenza infected pigs. This influenza also infected people and caused

---

Notes：Plague (*Yersinia pestis*)「ペスト」, penicillin「ペニシリン（抗生物質）」

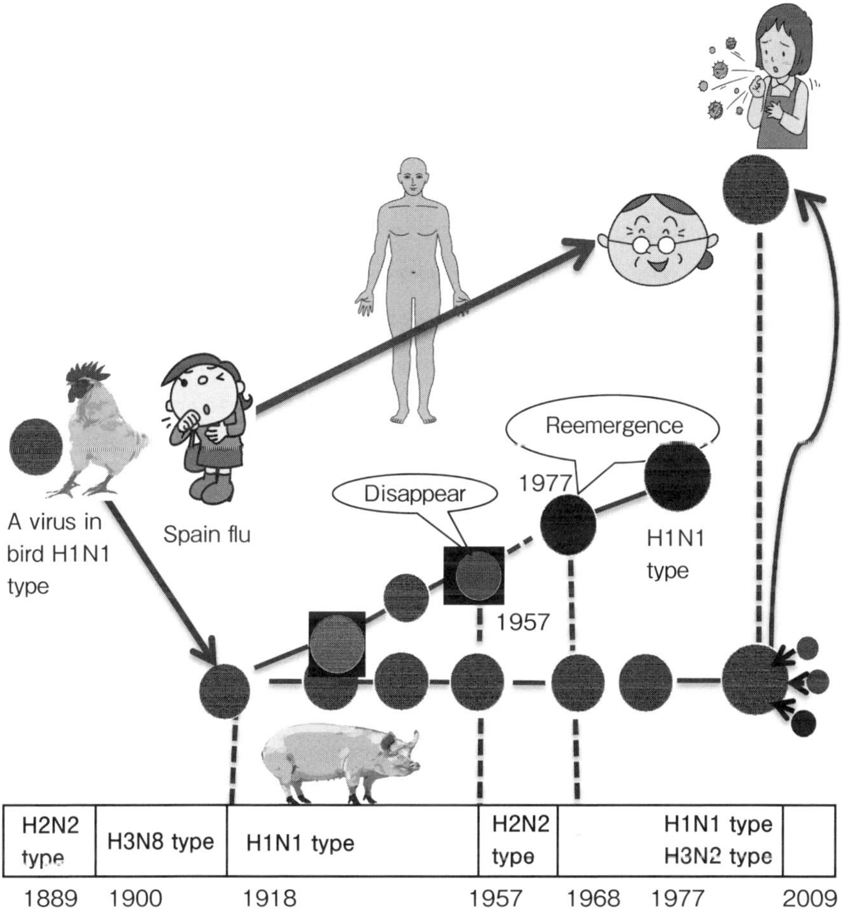

Figure 3.1 The transition in Influenza virus.
Elaborated from NIKKEI (18 Oct. 2009)

casualties. The infectiousness of this new-type influenza is high. In addition, owing to the proliferation of air travel, it can move across long distances in a short period and person-to-person infections occur within no time. In the course of a brief period, the WHO raised the worldwide pandemic alert level from Phase 3 to 4, further to Phase 5, and finally to the maximum of Phase 6 on June 12, 2009.

Most people infected by this swine influenza recover within a few days because of its weak virulence. As for the mortality rate, the WHO announced that relative to that of seasonal influenza (0.1 %), that of the new-type influenza was five times as high (0.5 %). This estimate is based on the mortality rate for the 1968 Hong Kong Influenza. However, a Harvard University study reported that there is no significant difference in mortality rates between the two influenzas. (Approximately 0.045 %). In either case, it is predicted that mortality rates will be higher for infants, those with underlying diseases, those who are pregnant, and the elderly.

What is an influenza virus? Influenza viruses are classified into types A, B, and C depending on their internal protein structures. Infection by a C-type virus triggers mild symptoms of a common cold. The B-type virus triggers a seasonal epidemic. Most problematic is the A-type virus. Although it may trigger a seasonal epidemic, at times it may trigger a global pandemic — the emergence of a new-type influenza. The influenzas that become worldwide pandemics are all based on A-type viruses. Past examples include the Spanish Influenza of 1918 (in which 4 million people died), the Asian Influenza of 1957 (in which 2 million people died), and the Hong Kong Influenza of 1968 (in which 1 million people died).

From the time it was first discovered, strange facts had been reported for the most recent new-type influenza. People born before 1918 have 50-60 % immunity. Those born in 1920 have only 20 % immunity. From this fact, it is thought that the Spanish Influenza, which became a pandemic and caused many deaths, did not actually die out but may have survived.

Note：common cold「一般的な風邪」

The structure of the influenza virus is one in which hemagglutinin (HA: a protein that determines virulence, with subtypes 1-16) and neuraminidase (NA) are inserted into a membrane called the envelope (**Figure 3.2**). There are sixteen types of HA and nine types of NA. These determine the protein structure of the virus, which in turn determines its virulence. For example, the currently circulating new-type influenza, which has been designated as the H1N1 subtype by taking the initial letters of each protein, falls into the same category as the Spanish Influenza and seasonal influenza A (Russian flu)

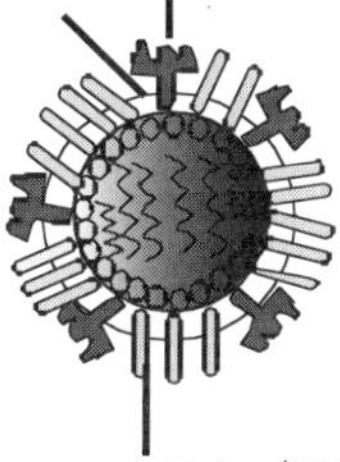

Figure 3.2 The swine influenza viruses and the structure detailed.

The virulence characteristics are determined by a combination of "shift" and "drift." First, "shift" refers to a large change in virulence caused by the re-aggregation of viral genes. In contrast, the change known as "drift" refers to mutations caused by small differences in the viral gene sequence. By using these two changes, influenza viruses are constantly changing their own virulence.

The true identity of the new-type influenza spreading at

Notes : hemagglutinin「ヘマグルチニン（抗原性糖タンパク質）」
neuraminidase「ノイラミニダーゼ：宿主細胞内で産生された複製ウイルスが細胞から遊離することを可能にする」

breakneck speed is a fusion of the present seasonal influenza A (Russian type, H1N1) — a mutated form of the Spanish Influenza, which was thought to have died out but for some reason survived — and a swine-originated new-type influenza.

How did such an event occur? The Spanish Influenza, which had spread like wildfire in the past, also infected swine. A normal influenza virus is constantly undergoing genetic mutation. However, given that none of the swine shipped elsewhere are immune, pressure on the virus to mutate was likely to have been low. For this reason, the properties of the Spanish Influenza were likely preserved in swine. This then, is the true nature of the birth of the new-type influenza virus.

Owing to the development of global transportation, influenza viruses will spread immediately across the world. Apart from the global expansion of the epidemic, the virulence is also considered a problem. Viral infections begin in the respiratory or gastrointestinal mucosa. If the virulence is low, the primary symptoms upon infection are coughing and diarrhea. However, if the virulence is high, the virus will infect the entire body; in addition to coughing and diarrhea, various symptoms associated with pneumonia and myocarditis will occur, causing multiple dysfunctions. For instance, if the virus reaches the brain via the blood, a symptom called influenza-associated encephalopathy may occur and lead to the loss of life within a short period. A virus with low virulence might mutate to a more virulent one is also likely possible.

This derives from the fact that the avian influenza virus is an RNA virus, that is, it mutates more easily than a DNA virus. Specifically, among the aforementioned HA proteins that determine

---

Note : gastrointestinal mucosa「胃腸粘膜」

virulence, viruses with only a single arginine sequencc are weakly virulent whereas those with repeated arginine sequences are highly virulent. Furthermore, as for the number of arginine sequences, influenzas that are RNA viruses can easily change their genetic sequence and may mutate into a virus of high virulence. A highly virulent influenza virus will cause an increased rate of mortality and become a super-influenza virus. No one knows when this mutation will occur.

The relation between people and viruses is ancient. Descriptions that hint of their circulation go back as far as 412 B.C. And as the population bccame denser with the development of civilization and distances between individual birds, swine, or people decreased, the number of people infected with the influenza virus increased. At the same time, the length of time between outbreaks is also decreasing. Ever since humans were chosen as one of its hosts by the influenza virus, the battle between people and viruses has been fated to continue eternally.

References

1. Books

1) Mike Davis：感染爆発—鳥インフルエンザの脅威．紀伊國屋書店（2006），ISBN978-4-314-01001-6
2) 岡田晴恵：感染症は世界史を動かす（ちくま新書）（2006），ISBN 978-4-480-06286-4
3) 山本太郎：新型インフルエンザ—世界がふるえる日—（岩波新書）（2006），ISBN 978-4-004-31035-8
4) 外岡立人：新型インフルエンザ・クライシス（岩波ブックレット）（2006），ISBN 978-4-000-09382-8

Notes：arginine「アルギニン（アミノ酸の１種）」
genetic sequence「遺伝子配列（塩基配列）」

5）堀本研子，河岡義裕：インフルエンザパンデミック—新型インフルエンザの謎に迫る—（ブルーバックス），講談社（2009），ISBN978-4-062-57647-5
6）NHK「最強ウイルス」プロジェクト 編：NHKスペシャル 最強ウイルス—新型インフルエンザの恐怖，日本放送出版協会（2008），ISBN978-4-140-81292-1
7）外岡立人：豚インフルエンザの真実—人間とパンデミックの果てなき戦い（幻冬社新書）（2009），ISBN978-4-344-98128-7
8）リチャード・E. ニュースタットほか著：豚インフルエンザ事件と政策決断—1976 起きなかった大流行，時事通信出版局（2009），ISBN978-4-788-70969-0

2. Journal

1）Cox N. J. and K. Subbarao：Global Epidemiology of Influenza；Past and Present Annual Review of Medicine **51**：pp.407-421（2000）

3. URL（2013年9月現在）

1）WHO 2013. Influenza update.
2）http://www.who.int/influenza/surveillance_monitoring/updates/latest_update_GIP_surveillance/en/

Exercises

1．つぎの1～70の語に対応する英単語または英熟語を本文から選び出し，発音しなさい（動詞は原形を記入しなさい）。

	Japanese	English		Japanese	English
1	インフルエンザ		8	蔓延（する）	
2	世界的流行（病）		9	抗生物質	
3	古代		10	不治の	
4	天然痘		11	ウイルス	
5	結核		12	下落（する）	
6	梅毒		13	効能	
7	コレラ		14	薬	

	Japanese	English		Japanese	English
15	新規の		36	生存する	
16	鳥類		37	構造	
17	死傷者		38	亜類型	
18	増殖する		39	挿入する	
19	上げる		40	膜	
20	毒性		41	外皮（封筒）	
21	推定する		42	循環する	
22	死亡率		43	指名する	
23	根底にある		44	組合わせ	
24	妊娠した		45	移行	
25	年配の		46	漂う	
26	仕分けする		47	再集成	
27	内部の，内面の		48	遺伝子	
28	タンパク質		49	突然変異	
29	誘起する		50	素性	
30	症状		51	非常に速い	
31	流行		52	融合	
32	出現，勃発		53	豚	
33	発券する		54	蔓延する	
34	免疫		55	野火	
35	事実		56	特性，財産	

	Japanese	English		Japanese	English
57	輸送		64	心筋炎	
58	拡張		65	機能不全	
59	呼吸の		66	脳	
60	胃腸の		67	脳症	
61	咳		68	説明，描写	
62	下痢		69	運命，因果	
63	肺炎		70	永遠の	

2. つぎの各文が本文の内容と一致するものには T（True），一致しないものには F（False）を，文末の（　　）に記入しなさい。

（1）It cannot be said that the history of mankind is a history of battles with infectious diseases.（　　）

（2）Viruses are new fears on mankind.（　　）

（3）Penicillin can cure any diseases.（　　）

（4）Symptoms of an influenza virus infection could lead to multiple dysfunctions.（　　）

（5）Outbreaks of the viral diseases are only recent problems in the world.（　　）

3. つぎの日本語の各文を（　　）の中の語を用いて英語の文にしなさい（必要があれば動詞を適切な形に変換しなさい）。

（1）感染病との戦いは，近い将来には終焉する。
(against, infectious)

（2）豚由来のインフルエンザは，人にも感染し，死亡者を出すこともあった。(swine, cause, casualty)

（3）感染性のあまりないウイルスでも，突然変異により毒性の高いものに変わることもあり得る。(virulence, mutate)

（4）RNA ウイルスは，DNA ウイルスよりもはるかに変異しやすい。(undergo, genetic change)

（5）ウイルスとヒトとの戦いは，永遠であるように運命づけられている。(fate, eternally)

4. つぎの各問いに英語で答えなさい。

（1）Would you like to be a clinical virologist?

（2）How can you predict see the epidemics of new diseases?

(3) How can an avian influenza virus become infectious to mankind?

(4) Do you think that one type of medicine can solve most of viral disease problems as penicillin is effective to many of bacterial diseases.

(5) The population became denser with the development of civilization and distances between individual birds, swine, or people decreased, the number of people infected with the influenza virus increased. Then, how should we do to keep our civilization?

Chapter 4
The Science of Seeing：A Perspective from the World of Neuroscience

Whenever we recognize an object, we are using our five senses of sight, hearing, touch, smell, and taste. Each of these senses is undoubtedly important. Of these, the acquisition of information by vision in particular is most important. Of the five senses, vision accounts for about 76 % of the total information involved in recognizing an object, followed by 11 % for hearing, 5 % for taste, 5 % for touch, and 3 % for smell. Although each function is important, impaired vision in particular is very pronounced.

The causes of visual impairment can be divided broadly into two types: acquired and genetic or constitutive causes. Worldwide, blindness due to cataracts accounts for 51 % of visual impairments. In subsequent order of prevalence, other causes include glaucoma (12.3 %), age-related macular degeneration (8.7 %), corneal opacity (5.1 %), diabetic retinopathy (4.8 %), childhood blindness (3.9 %), trachoma (caused by parasitic infection; 3.6 %), and onchocerciasis (caused by parasitic infection; 0.8 %). In recent years, lens implants have become easy to perform in developed countries owing to technological progress, and the prevalence of cataracts is declining. Meanwhile, in developing countries, cataracts account for about half of all blindness. Accidents are the most common acquired causes of blindness or weak eyesight.

Notes：corneal opacity「角膜翳，角膜の混濁異常」

diabetic retinopathy「糖尿病性網膜症」

onchocerciasis：river blindness and Robles disease「センチュウの一種である *Onchocerca volvulus* による感染で発症する」

Various efforts are being made to treat blindness. Corneal transplantation is one of the procedures used after a person becomes blind. In Japan, few people register to donate to so-called "eye banks." Therefore, many people go overseas for transplants. There is one major problem with corneal transplantation — a problem that exists especially for people who have lost their sight in infancy by accident, and in which the neuroscience of seeing is highly relevant.

In people who become blind in infancy, the neurons related to the visual sense that is no longer used are reassigned, for example, to increase their sensitivity to other senses. This is called "neuroplasticity." When we see an object, we do not in fact see what is projected onto the eye as is. The image projected onto the retina deep behind the lens is collected by photoreceptor cells and transformed into electrical signals. Only then is it possible to recognize an object. In other words, if there is any defect in the optic nerve, objects may appear to be invisible, partially invisible, or deformed even if the eye itself is perfectly normal.

For instance, a person blind since infancy as the result of an accident, despite having a successful corneal transplant, will still need subsequent rehabilitation and considerable time before he or she can accurately "see" an object for what it is.

In fact, for an object to be seen accurately, a complex mechanism requiring the following three elements is involved:

1) The ability to perceive faces
2) The ability to perceive depth
3) The ability to recognize objects

1) The ability to perceive faces

First, a healthy person is quite naturally endowed with the ability to perceive faces. However, for those who have been blind

---

Note : photoreceptor「視細胞，植物では光受容体」

for a long time, this task is in fact very difficult. It turns out that human faces are not too different from one another. For example, both a healthy person and a person that was once visually impaired would probably be able to distinguish an elephant from a rabbit, which differ greatly in terms of their size and facial parts.

In comparison, the differences between human faces are quite small. People that have long been blind cannot distinguish between a face in which the eyes and mouth are in the proper positions from one in which they are rotated 180 degrees, even when they know the differences between nose, eyes, eyebrows, and mouth (**Figure 4**). This may seem very surprising. In fact, people have acquired this experience over a long period through learning.

2) **The ability to perceive depth**

The complexity of this skill is similar to that of the ability described above. The depth of an object is actually determined by the size of the object and how far it is from the observer. We are likely to know the distance from an object for the first time only after actually experiencing it (first by walking the distance ourselves). By repeating this over and over, we become able to tell how far we are, for example, from a pair of glasses on a desk (even if they are unreal), by recalling past memories. Specifically, we can perceive depth only when we imagine the size of the glasses as well as that of the desk.

3) **The ability to recognize objects**

This skill is quite similar to the above-described ability to perceive depth. With the ability to see an object in three dimensions, estimating the angle at which the object is projected becomes possible whatever the angle. This ability is said to take

A subject can be recognized from different view angles with a normal sight.

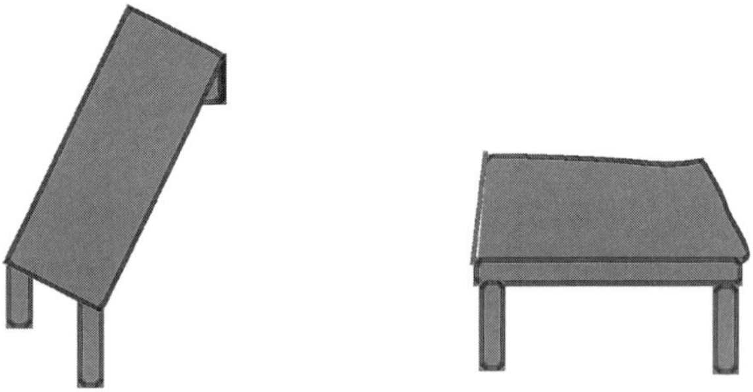

These two tables are same size, and a person a with normal view sight can be confused.

A person with normal sight can recognized that this woman is the same person.

Figure 4 The some illusions occur in a normal view sight, while not occur in a person with a long blindness.

several years to acquire.

Only by integrating these three abilities can a person “see” an object. In other words, it is the brain, not the eyes, that ultimately sees an object. An eye is merely a machine that transfers the information projected onto its surface to the optic nerves. We see with our brains. Our brains have approximately 100 billion neurons (nerve cells), each of which plays a role similar to that of an electrical signal, making corrections where necessary and sending the resulting electrical signal to the next neuron. Individual neurons are connected to other neurons to form a network of neurons. The act of seeing becomes possible only when the “seeing” network is engaged. It goes without saying that separate networks of neurons are engaged for other acts such as smelling or moving one’s legs.

Owing to recent research, the properties of these neurons are beginning to be revealed. For example, if a person loses the ability to see or perform a particular movement, and electrical signals stop arriving at a neuron, its function changes to a different one as the function that had been expected is discarded. This so-called “plastic” property becomes detrimental to restoring vision when the person becomes an adult. In reality, new neuronal connections are less likely to form in adults. Meanwhile, there are reports that vision recovers considerably by continuing rehabilitation after corneal transplantation. Neuroscience is the most challenging field remaining among the various areas of medicine. In addition to corneal transplantation, research on embryonic stem (ES) cells and induced pluripotent stem (iPS) cells, is advancing rapidly in the area of vision. The day when a blinded person can recover her/his vision is on the horizon.

Notes : embryonic stem (ES) cells : See Chapter 1

induced pluripotent stem (iPS) cells : See Chapter 1

References

1. Books

1）大森　聡：クリアな瞳―角膜移植による視力向上，ウィズダムブック（2005），ISBN978-4-901-34716-7
2）ロバート・カーソン：46年目の光―視力を取り戻した男の奇跡の人生―エヌティティ出版（2009），ISBN978-4-757-15060-7
3）玉井　信，水流忠彦：眼組織移植と免疫．（NEW MOOK眼科（3））金原出版（2003），ISBN978-4-307-63303-1
4）坪田一男：不可能を可能にする視力再生の科学．（PHPサイエンス・ワールド新書）（2010），ISBN 978-4-569-77785-6
5）坪田一男，島崎　潤，榛村重人：角膜移植ガイダンス―適応から術後管理まで，南江堂（2002），ISBN978-4-524-22242-1
6）山中伸弥，中内啓光 編：再生医療へ進む最先端の幹細胞研究，実験医学増刊 **26**-5，羊土社（2008），ISBN978-4-758-10289-6
7）V. S. ラマチャンドラン，D. ロジャース ラマチャンドラン：知覚は幻―ラマチャンドランが語る錯覚の脳科学，別冊 日経サイエンス 174（2010），ISBN 978-4-532-51174-6
8）クリストファー・チャブリス，ダニエル・シモンズ：錯覚の科学，文藝春秋（2011），ISBN978-4-163-73670-9

2. Journal

1）M. Eiraku, et al. : Nature **472**, pp.51-52（2011）
2）F. Osakada, et al. : Nature Biotechnology **26**, pp.215-224（2008）

3. URL, Movies（2013年9月現在）

1）http://www.who.int/blindness/causes/priority/en/index1.html
2）http://nature.asia/nd06-nv-video1
3）http://nature.asia/nd06-nv-video2
4）http://nature.asia/nd06-nv-video3
5）http://nature.asia/nd06-nv-video4
6）http://nature.asia/nd06-nv-video5
7）http://nature.asia/nd06-nv-video6
8）http://nature.asia/nd06-nv-video7
9）http://nature.asia/nd06-nv-video8

Exercises

1. つぎの 1 ～ 44 の語に対応する英単語または英熟語を本文から選び出し，発音しなさい（動詞は原形を記入しなさい）。

	Japanese	English		Japanese	English
1	感覚		19	減少，低下	
2	神経科学		20	移植	
3	獲得		21	寄付する	
4	視覚		22	神経細胞	
5	機能		23	神経可塑性	
6	障害のある		24	網膜	
7	宣告する		25	欠陥	
8	構成的な		26	視力の	
9	盲目		27	見えない	
10	白内障		28	変形させる	
11	流行		29	復帰	
12	緑内障		30	知覚する	
13	筋肉の		31	授ける	
14	退化		32	識別する	
15	トラコーマ		33	回転する	
16	寄生の		34	眉	
17	レンズ（水晶体）		35	技能	
18	移植する		36	次元	

	Japanese	English		Japanese	English
37	従事する		41	有害な	
38	特性，財産		42	復元する	
39	廃棄する		43	接続	
40	プラスチックの		44	回復する	

2. つぎの各文が本文の内容と一致するものにはT（True），一致しないものにはF（False）を，文末の（　　）に記入しなさい。

（1）Humans have sixth sense proven definitely by science.（　　）

（2）Five senses of humans are undoubtedly important.（　　）

（3）In developing countries, cataracts account for a quarter of all blindness.（　　）

（4）In Japan, few people register to donate to so-called “eye banks.” Therefore, many people go overseas for transplants.（　　）

（5）A healthy person is not naturally endowed with the ability to perceive faces.（　　）

3. つぎの日本語の各文を（　　）の中の語を用いて英語の文にしなさい（必要があれば動詞を適切な形に変換しなさい）。

（1）視覚障害の原因は大きくは2種類に分別される。（cause, divide, impairment）

（2）近年，技術発展により，先進国では水晶体の移植が容易になってきている。（lens, perform, facile）

（3）眼球が完璧であっても，視神経になんらかの異常があると，物体は，不可視，部分的に見えない，あるいは変形して見える。（optic, object, deform）

（4）個々の神経細胞は相互に連結し神経細胞の連絡網を形成する。（individual neuron, form, connect）

（5）角膜移植の後，継続した回復鍛錬をすると飛躍的に視力が回復することが報告されている。（vision, recover, rehabilitation, corneal）

4. つぎの各問いに英語で答えなさい。

（1）How do you think about the sixth sense?

（2）How do you classify the causes of visual impairment?

（3）Owing to recent research, the properties of the neurons are beginning to be revealed. How would you like to engage in such research efforts?

（4）Research on embryonic stem（ES）cells and induced pluripotent stem（iPS）cells, is advancing rapidly in the area of vision. Do you agree to test the research outcome immediately to the blind patients?

（5）For an object to be seen accurately, What are the key elements?

Chapter 5
Did You Sleep Well? :
The Increasing Severity of Sleep Disorders

Recently, many people are dissatisfied with their sleep. One out of every five people is said to have some sleeping disorder. The content of the complaints differ greatly, such as sleeping extremely short hours, not sleeping deeply enough, and waking too early. Indeed, sleep is a complexity of phenomena (**Figure 5.1**). This phenomenon has led to a new field called sleep science, in which various studies are being conducted to reclaim a comfortable sleep. Here we introduce examples of a number of sleep disorders.

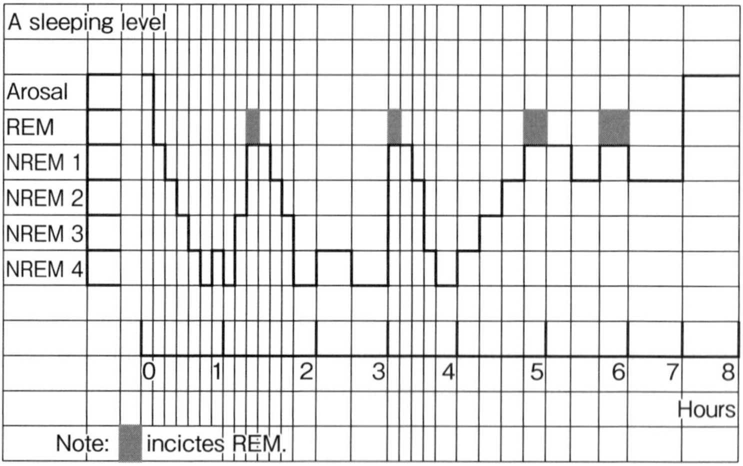

Figure 5.1 A sleeping rythm of a human being.

False memory

By far the most common complaint is shortage of sleep. If a person's lifestyle becomes increasingly night-oriented and the time spent awake increases, then the hours of sleep will naturally decrease. This shortage of sleep is causing new problems. Everyone has probably experienced that when we don't get enough sleep, our heads become unclear, and we become more forgetful. Why then does our memory decline when we become short of sleep?

Memory is thought to be stabilized when we are asleep at night. An experiment was conducted with two groups. Before the experiment, the subjects were asked to memorize four simple terms that were related. The first group got an adequate night's sleep, whereas the second group did not sleep at all. The subjects were then asked to recall the memorized words. A difference was found between the two groups, with the group that had not slept showing significantly more incorrect answers.

This false memory occurs frequently in groups that do not get adequate sleep during the night. Furthermore, when the same memory test was performed again for the second group (the group that had not slept), but this time after they had adequate sleep, the erroneous answers declined and did not differ from those of the first group (the group that had adequately slept). Their false memories had thus returned to normal.

When we consider these results scientifically, memory itself is likely formed regardless of whether or not the subjects slept. However, there was a problem in the memory retrieval component. In the group that gave incorrect answers, a false memory is retrieved from memory storage. The memory is apparently stabilized the moment it is awakened. This appears to occur because the prefrontal cortex, which is involved memory, is not adequately functioning owing to inadequate sleep.

Then what about the effect of caffeine, which helps to keep us

---

Note: prefrontal cortex「前頭葉前皮質」

awake? False memory has been confirmed to decrease by 10 % when caffeine is ingested, even if subjects had not adequately slept. Caffeine is known to act on the prefrontal cortex of the brain. The site of this action in the prefrontal cortex is also the place that distinguishes between events that actually occurred and events that were only considered. This becomes deeply related to the issue of whether a witness accurately remembers an event when testifying in court. How this relates to the quality rather than the quantity of sleep remains unknown, however, and other factors may also turn out to be involved. Research on the mechanisms by which false memories are produced has only just begun. If we should ever be placed on the witness stand with substantial lack of sleep, we may be better off going to court after drinking some coffee.

Narcolepsy

"Narcolepsy" may be translated as "napping disorder" (**Figure 5.2**). A movie based on this topic, "The Cabinet of Dr. Caligari", has also been released. The first thing that one recalls is the story, in which a sleeping killer appears and kills his companions one after another. The term narcolepsy is a disease name that links "narco" "numbness" or "stupor" and "lepsy" (attack). As its name implies, narcolepsy is indeed a "sleep attack."

The characteristic feature of narcolepsy is that sleepiness strikes suddenly. Unlike the sleepiness you feel when listening to a boring story drag on or during a class, the narcoleptic immediately ends up in a deep sleep. And as soon as sleep begins, the narcoleptic frequently experiences terrifying dreams — nightmares no one would want to see: for example, dreams about being chased by a person or

---

Note：narcolepsy「睡眠発作；日中において急に眠気の発作が生じる脳疾患」

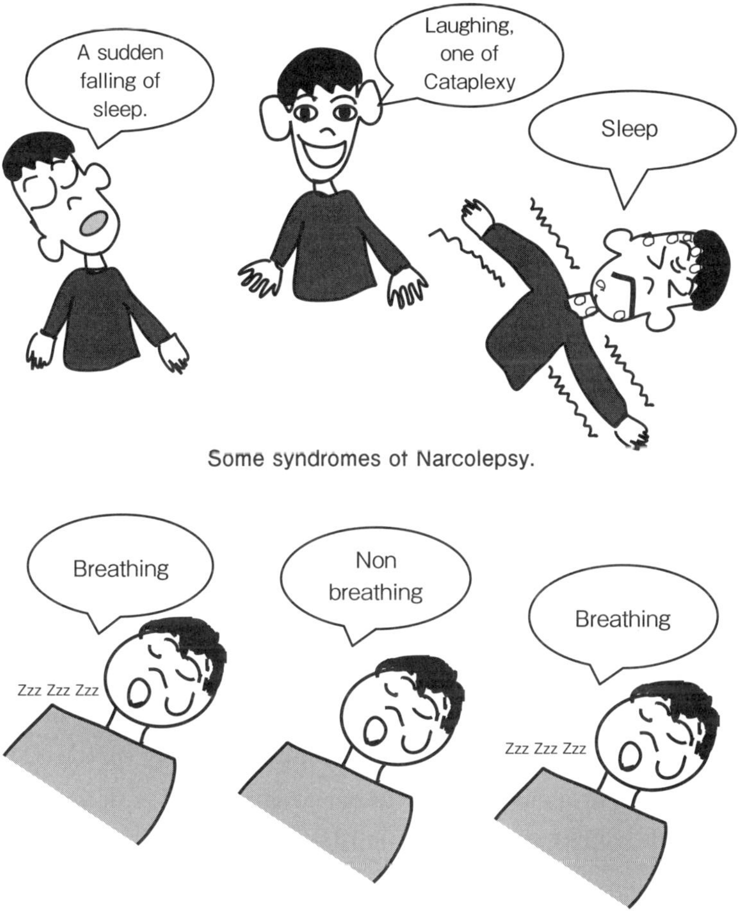

Figure 5.2　Examples of sleep disorders: Narcolepsy and Sleep Apnea Syndrome.

being strangled. This is closely related to the type of sleep.

A normal adult first goes through cycles of initial “non-REM sleep” followed next by “REM sleep (Figure 5.1).” The brain and body are most at rest during non-REM sleep. In contrast, although the

---

Note：REM sleep「REM は rapid eye movement の略，睡眠は 2 相に分けられ REM 睡眠は，このうち脳波があたかも覚醒時のような状態になっている」

body is resting during REM sleep, the head is working as when the subject is awake (the acronym REM comes from Rapid Eye Movement). In fact, during REM sleep, the eyes move around as if the subject were awake. In contrast, the term "non-REM sleep" is used to indicate a type of sleep that differs from REM sleep. The reason why a subject awakened during REM sleep can get up with a clear head is because the head is functioning as it does when awake. Normally, "non-REM sleep" and "REM sleep" repeat every 90 minutes or so. However, in the case of narcolepsy, sleep starts with REM sleep rather than non-REM sleep. For this reason, the head functions as it does during wakefulness, and thus it dreams. This is called sleep-onset REM.

In addition, the narcoleptic may also fall into a state of sleep paralysis, or in some cases begin to show symptoms called "cataplexy" even when awake — either falling to the ground, losing the ability to exert strength, or falling into a state of weakness.

The narcoleptic also has problems with night-time sleep. Contrary to what "napping disorder" implies, sleep disruptions occur frequently, with subjects often being able to take only scattered bouts of sleep. Whether a subject is or is not a narcoleptic is determined by verifying the combination of each of these symptoms.

Narcolepsy is said by some to develop in 1 out of every 1 000–1 500 people, and the mechanisms underlying its onset are also being studied. First, an examination of genes has revealed that narcoleptics possess the DRB1*1501-type of HLA (human leukocyte antigen) gene. However, not all who possess this gene necessarily develop narcolepsy; only when other factors such as stress overlap does narcolepsy finally develop. In a twin study as well, the probability of both developing the disease was not 100 % and less than 50 % for

---

Note : cataplexy「脱力発作」

each.

Orexin levels in the cerebrospinal fluid were also found to be low for narcoleptics. Orexin plays a role in facilitating the sleep-wake cycle. Currently, low levels of orexin production are considered as a potential cause of narcolepsy.

The first step in the treatment of narcolepsy starts with the patient fully understanding that he or she is a narcoleptic. It is also necessary for those surrounding the patient to understand that narcolepsy is not laziness and that treatment is required. As a treatment method, the first thing is to make it a habit of taking a nap to clear the head (based on the fact that the head is very clear after a narcoleptic event). The second is to avoid heavy drinking or irregular lifestyles and not to accumulate stress. Another way, depending on the state of the disease, is to take a psychostimulant during the daytime and a sleeping pill at night. For nearly every case of narcolepsy, the onset occurs in the teens to 30s, and the symptoms often become milder with age. However, the reason for this remains unclear.

Sleep apnea syndrome (*SAS*)

An incident occurred in which a *Shinkansen* (bullet train) did not stop at a station. In the course of investigating the cause, it was revealed that the driver had sleep apnea syndrome (SAS) and that during the time in question, he had been struck by a bout of sleepiness and failed to stop the train. The term SAS became widely known.

A characteristic feature of SAS from the perspective of sleep medicine is first that breathing stops for 10 seconds or more (apnea). As a result, patients often do not get adequate sleep during the night,

---

Notes：orexin「神経ペプチドの一種」, cerebrospinal fluid「脳脊髄液」
psychostimulant「覚醒剤」, sleep apnea syndrome「睡眠時無呼吸症候群」

and experience bouts of daytime sleepiness. Other characteristics include lack of concentration, heavy-headedness or discomfort on awakening, depressive feelings, and impotence. Different types of SAS can be classified. In the obstructive type (OSAS), the abdominal and thoracic areas continue to move but oronasal air flow ceases. In the central type (CSAS), oronasal air flow ceases, and neither abdominal area nor thoracic area moves. In the latter case, it is thought that the respiratory center in the brain does not send commands. In addition, there are mixed types involving a mixture of the obstructive and central types. The obstructive type is by far the most common.

Obstructive sleep apnea (OSA) is diagnosed when the apnea/hypopnea index is at least 5, in other words, if there are 5 or more episodes of sleep apnea in 1 hour. The rates of affliction are 3.3 % for males and 0.5 % for females, with males accounting for a large majority. A characteristic of OSAS, the most common SAS, is a thickening of the neck from obesity. This is caused by a softening of the airway walls due to the sagging of fat, which in turn obstructs the airway. Additionally, close attention should be paid to those with a small lower jawbone.

In terms of the skeletal framework of the jaw, the airway tends to be easily obstructed in those whose lower jaw is pushed backwards, or whose upper jaw protrudes forward. Those in whom the tonsils at the back of the throat are enlarged are also at risk. In particular, enlarged tonsils is the number one cause of OSAS in children.

SAS is also likely to cause complications with other ailments, such as hypertension, arrhythmia, ischemic heart disease, and diabetes. In the worst case, it may also cause sudden death.

The treatment considered most effective is weight reduction. As

---

Notes：OSAS「閉塞性睡眠時無呼吸症候群」, ischemic heart disease「虚血性心疾患」

a result, the throat is no longer obstructed, and the probability of improvement is considerable. In cases where there is still no improvement, the airway is secured by wearing a mouthpiece to prevent the lower jaw from sagging while sleeping. If no improvement is seen even with these methods, a Continuous Positive Airway Pressure（"CPAP"）system can be used. Although there is some initial discomfort, patients gradually become accustomed and in many cases are finally able to get a good night's sleep.

Humans spend one third of their lives sleeping. One could say that this shows the extent to which sleep is an important indicator of health. If your anxieties concerning sleep grow, it may be worth visiting the outpatient clinic of a hospital sleep unit for a checkup.

References

1. Books

1）上里一郎，白川修一郎：睡眠とメンタルヘルス―睡眠科学への理解を深める（シリーズ こころとからだの処方箋）ゆまに書房（2006），ISBN978-4-843-31820-5
2）内田　直：好きになる睡眠医学，講談社（2006），ISBN978-4-061-54162-7
3）太田龍朗：睡眠障害ガイドブック―治療とケア，弘文堂（2006），ISBN978-4-335-65124-3
4）北浜邦夫：脳と睡眠，朝倉書店（2009），ISBN978-4-254-10215-4
5）櫻井　武：睡眠の科学―なぜ眠るのかなぜ目覚めるのか（ブルーバックス）講談社，（2010），ISBN978-4-062-57705-2
6）成井浩司：睡眠時無呼吸症候群がわかる本．法研（2005），ISBN978-4-879-54567-1
7）堀　忠雄：快適睡眠学の勧め，（岩波新書）（2000），ISBN978-4-004-30683-2
8）本多　裕：ナルコレプシーの研究―知られざる睡眠障害の謎（Hot-nonfiction），悠飛社（2002），ISBN978-4-860-30018-0
9）松浦雅人 編：睡眠検査学の基礎と臨床，新興医学出版社（2009），ISBN978-4-880-02692-3

Note：(nasal) continuous Positive Airway Pressure「(経鼻的）持続陽圧呼吸療法」

2. Journal

1) Y. Harrison and JA Horne：Sleep Loss and Temporal Memory. Quarterly J. Exp. Psychology **53** (A)：pp.271–279 (2000)
2) Payne, et al.：The Role of Sleep in False Memory Formation. Neurobiol. Learn. Mem. **92**, pp.327–334 (2009)
3) B. Straube：An Overview of the Neuro-Cognitive Processes Involved in the Encoding, Consolidation, and Retrieval of True and False Memories. Behavioral and Brain Functions 8：35

3. URL (2013年9月現在)

1) http://www.nature.com/news/2008/080714/full/news.2008.953.html

Exercises

1. つぎの1～58の語に対応する英単語または英熟語を本文から選び出し，発音しなさい（動詞は原形を記入しなさい）。

	Japanese	English		Japanese	English
1	障害		12	摂取する	
2	現象		13	作用する	
3	取り戻す		14	証人	
4	安楽な		15	証言する	
5	夜型の		16	実質の	
6	起床した		17	昼寝，居眠り	
7	安定させる		18	放つ	
8	記憶する		19	含蓄する	
9	遂行する		20	特徴，容貌	
10	誤った		21	恐れさせる	
11	回復，挽回		22	窒息する	

	Japanese	English		Japanese	English
23	覚醒		41	無呼吸	
24	麻痺		42	低呼吸	
25	崩壊，中断		43	苦悩	
26	散発的な		44	下がる，たわむ	
27	発作，ひとしきり		45	顎骨	
28	基礎となる		46	出来事	
29	検査		47	突き出る	
30	促進する		48	骨格の	
31	症候群		49	扁桃	
32	集中，濃度		50	咽喉	
33	うつ病の，憂鬱(うつ)な		51	疾病	
34	妨害する,閉塞性の		52	高血圧	
35	腹部の		53	不整脈	
36	胸部の		54	マウスピース	
37	口腔鼻		55	指標	
38	無気力		56	心配	
39	呼吸の		57	外来医療	
40	診断する		58	健康診断，照合	

2. つぎの各文が本文の内容と一致するものにはT(True)，一致しないものにはF(False）を，文末の（　　）に記入しなさい。

（1）One out of every ten people is said to have some sleeping disorder.（　　）

（2）This false memory occurs frequently in groups that do not get enough sleep during the night.（　　）

（3）The memory is apparently stabilized at the moment it is in sleep.（　　）

（4）Caffeine is known to act on the prefrontal cortex of the brain.（　　）

（5）A normal adult first goes through cycles of initial "REM sleep" followed next by "non-REM sleep".（　　）

3. つぎの日本語の各文を（　　）の中の語を用いて英語の文にしなさい（必要があれば動詞を適切な形に変換しなさい）。

（1）試験の前に，被験者は関連する4事項について記憶するように求められた。(subject, memorize, terms)

（2）夜間に十分睡眠しなかったグループでは，間違った記憶が頻繁に起こった。(false, adequate, frequently)

（3）OSASの特徴は肥満による頸部の肥大である。(thickening, obesity)

（4）OSAは無呼吸低呼吸指数が5以上である場合に診断される。(diagnose, apnea/hypopnea index)

（５）睡眠発作のほとんどすべてのケースにおいて，その徴候は 10 代から 30 代のうちに現れ，そしてその症状は，しばしば年齢とともに軽くなっていく。（narcolepsy, symptom, onset）

4. つぎの各問いに英語で答えなさい。

（１）What are steps of the treatment of narcolepsy?

（２）What are characteristic features of SAS from the perspective of sleep medicine ?

（３）With what illness sleep apnea syndrome（SAS）would lead to more complication?

（４）What would be the best treatment on SAS?

（５）Why do Humans spend one third of their lives sleeping?

Chapter 6
Japan's National Disease : Hay Fever

Nowadays, hay fever has replaced tuberculosis as "Japan's national disease." Among those with hay fever, an overwhelming number are allergic to Japanese cedar (*Cryptomeria japonica*). Come early spring when the pollen from Japanese cedars drifts down, towns are filled with people wearing face masks. Around this time, many people opt to dry their laundry inside. One study has estimated that the incidence of this hay fever is 16.2 % (2006) of the population. The main symptoms are sneezing, runny nose, nasal congestion, and itchy eyes (primary symptoms). (**Figure 6.1**)

Figure 6.1 A lot of pollens scatter over the planet at the early spring.

If these symptoms are severe, they may lead to throat disorders due to mouth breathing, coughing or phlegm, and symptoms resembling asthma. Alternatively, eye symptoms may become severe, with people developing conjunctivitis or suffering from the feeling that they

---

Notes : hay fever「花粉症 (allergic rhinitis)」, nasal congestion「鼻づまり」
itchy eyes「目のかゆみ」, mouth breathing「口呼吸」

have something in their eyes. Sufferers may also experience itchiness deep in the ear, nose bleeds caused by excessive nose-blowing, and thickening nasal discharge leading to symptoms resembling sinusitis. Some people develop bronchitis when nasal discharge enters the airway and causes inflammation. Others will develop headaches, gastrointestinal symptoms, such as diarrhea, nausea, and stomachache, and dermatitis caused by pollen. Additionally, some may complain of lack of concentration due to insomnia, irritability, and lack of appetite (secondary symptoms). (**Figure 6.2**)

Figure 6.2 Many symptoms occur with the allergic problems.

Hay fever is, as might be expected, indirectly caused by plants that produce pollen. Particularly common in Japan is hay fever due to Japanese cedars. The other characteristic of hay fever in Japan is that the pollen mainly responsible is carried by the wind (i.e., the flowers are wind-pollinated). Compared to insect-pollinated flowers, the pollen of Japanese cedar is tiny (20 μm in diameter) and thus travels longer distances. When the pollen season begins, the sky turns also yellow as when an outbreak of yellow sand occurs in Asian continent,

Notes : nose bleed「鼻血」, nose-blowing「鼻をかむこと」, nasal discharge「鼻汁」 wind-pollinated「風媒」, insect-pollinated「虫媒」

and the sands blown from the inland China. The pollen from Japanese cedars that bloom in the north Kanto region can travel more than 100 km. The pollen is carried to the south Kanto region and causes many to suffer.

Our noses start to itch when we inhale Japanese cedar pollen because of the antigen-antibody reaction we possess as humans. When we inhale pollen through our nose, the allergen — the substance that triggers the allergy — dissolves out of the pollen in the mucous membranes lining the nose. Antibodies are produced in the body to battle these antigens. When these antibodies capture the antigens, they secrete histamine, which stimulates the mucous membranes of lining the nose and leads to continuous nasal discharge. Japanese cedar pollen consists of various components. Recent studies suggest that rather than the pollen itself, it is its outer membrane that triggers the itchiness.

This antigen-antibody reaction is a welcome function originally designed to protect humans by discharging foreign objects to prevent them from adversely affecting the body. However, when these antibodies exceed a certain level, nasal discharge may become continuous and the eyes may become itchy. Hence, beyond a certain level, many people suddenly develop hay fever in adulthood despite not having it during childhood. An allergic reaction is very uncomfortable. It is an immune reaction, and the underlying system is the same as that in which a person once having contracted mumps or measles will never do so again.

Hay fever is classified as a Type I (immediate-type) allergic reaction. Also included in this group are bronchial asthma, atopic

---

Notes : antigen-antibody reaction「抗原抗体反応」, mucous membrane「粘膜」
atopic dermatitis「アトピー性皮膚炎」

dermatitis, and gastrointestinal allergies. In people with Type I allergic reactions, known as those with an atopic disposition, an antibody called immunoglobulin E (IgE) is produced when an allergen enters the body. This IgE antibody binds with the receptors on the fat cells of skin and mucosa, and triggers the allergic reaction.

Why does hay fever develop? There was speculation that the constitution for hay fever might be inherited. Upon analyzing a group of people with cedar pollen allergy, a gene called HLA (human leukocyte antigen) was found to be involved. It was found that cedar pollen allergy is more likely to occur in individuals who inherited two recessive alleles of HLA, and resistance to cedar pollen was more likely to occur in individuals with two dominant alleles. Is everything then determined by genes alone? Actually, there are many cases in which cedar pollen allergy does not develop despite individuals being recessive for the gene. What else is involved?

Environmental factors are also important for the development of hay fever. Specific factors include exposure to the allergen, air pollution, infections, living environment, and eating habits. Additionally, physiological factors are also associated, where age, sex hormones, and autonomic nerves are equally important. Hay fever is thus a multi-factor disease in which genetic, environmental, and physiological factors mutually interact.

The pollen that causes hay fever (foremost being Japanese cedar) circulates all year. In fact, hay fever due to Japanese cedar alone develops in more than 16 % of all who suffer from pollinosis. In addition, more than 40 types of pollen including Japanese cypress, ragweed (*Ambrosia spp.*), and mugwort (*Artemisia vulgaris*) have

---

Notes : atopic disposition「アトピー症状」, immunoglobulin E「免疫グロブリン E」
human leukocyte antigen「ヒト白血球抗原」

been reported to cause hay fever. Unfortunately, we will always inhale pollen unless we continue to stay indoors.

What therapies are available when should you happen to develop hay fever? The easiest method is to avoid direct inhalation of pollen as much as possible, such as by wearing a mask. To the best possible extent, brush off pollen outside after returning home and be careful not to bring in any indoors. We should also dry our clothes indoors during periods of pollen circulation and as much as possible wear pollen-free clothing. We could also look up the amount of pollen in the air for any particular day. By using "Hanako", the Meteorological Agency's Pollen Circulation Prediction System, we can know the amount of pollen circulating for Japanese cedar and Japanese cypress in real time. This information can help us take measures to prevent pollen inhalation.

If the symptoms still do not subside, then we need to resort to therapeutic drugs for hay fever. Treatments to control hay fever fall largely into six categories: 1) chemical mediator release inhibitors, 2) chemical mediator receptor antagonists, 3) Th2 cytokine blockers, 4) steroids, 5) autonomic agonists, and 6) others. The majority of therapeutic drugs are those that suppress the secretion of histamine, which causes the symptoms. Although these ameliorate the symptoms, one method fundamentally improves the atopic disposition itself. This is what is known as desensitization therapy. The basic idea of this treatment is to gradually acclimatize the body to the allergen. Highly diluted allergen (100 000:1) is first injected, after which the concentration is gradually increased. The effects are consistent, although the treatment period is long, ranging from 1 to 2 years, or

---

Note：cytokine「サイトカイン；免疫反応において抗原がリンパ球に結合した際リンパ球から分泌されるタンパク質」

even longer, depending on the person. Other available methods include surgery or laser treatment to open the nasal passages for easier breathing. However, the effects are short-lasting in many cases because of the great ability of the mucous membranes to repair themselves (**Figure 6.3**).

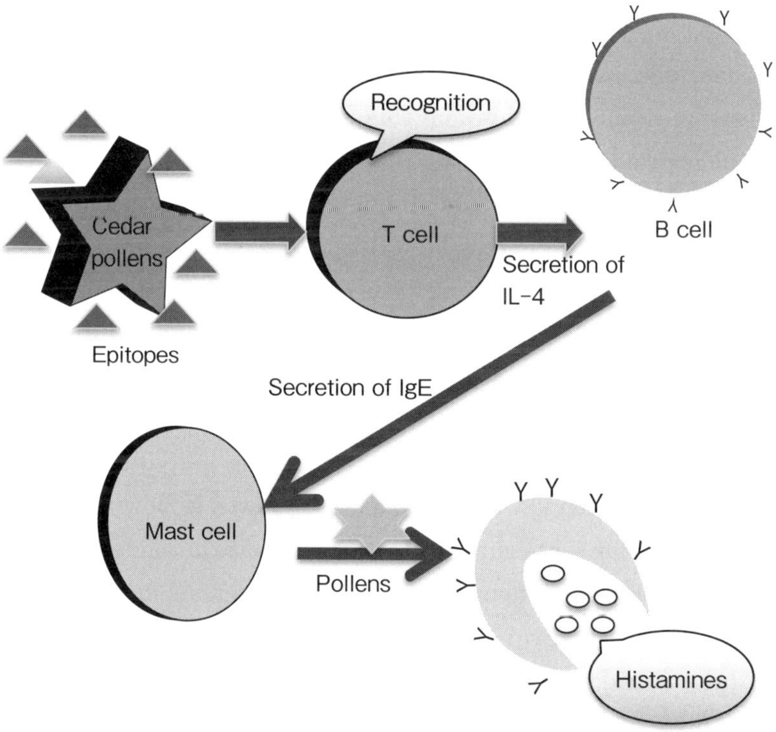

Figure 6.3 A step of allergic occurrence.

Although hay fever is affected by the weather in a particular year, current trends indicate that the incidence is increasing yearly. One calculation even suggests that 40 % of citizens will have hay fever by around 2025. In addition, people engaged in forestry are aging, and areas where tree thinning cannot be performed are likely to increase, leading to the release of ever more Japanese cedar pollen in the future. And intertwined with this is the problem of global warming, adding fuel to the fire. The ensuing rise in temperature and hours of sunshine ensure advantageous conditions for the Japanese cedar. It is almost certain that the amount of pollen will rapidly increase and that the incidence rate of hay fever will rise.

Must we then simply look on in blank amazement at these conditions, and spend our days fearing when we will develop hay fever ourselves? A number of views on this problem have been worked out from a biological approach.

Efforts are being made to eliminate pollen, in particular that of Japanese cedar. At the Forest Tree Breeding Center, a unique (i.e., a mutant) individual (male sterile) was discovered among male flowers of the Japanese cedar. This individual, while bearing male flowers, does not mature; hence, it does not produce pollen. The problem lies in how to propagate it. Naturally, as the male flower does not produce pollen, it cannot produce seeds. Therefore, this individual must be propagated from cuttings. However, as tree growth is slow, it will take several decades for these cuttings to reach the size of the original tree. There are also attempts to propagate these individuals for the purpose of cell culture. However, they must grow normally; in addition, individuals suited to the purposes of forestry must be confirmed (**Figure 6.4**).

An "edible vaccine" is under development as well, which at present is very close to practical implementation. This product utilizes a desensitization method and was developed by genetically engineering rice, the main staple of Japan. Specifically, the substance by which an allergen recognizes an antibody (the so-called epitope,

---

Note : epitope「抗体が認識する抗原との結合部分」

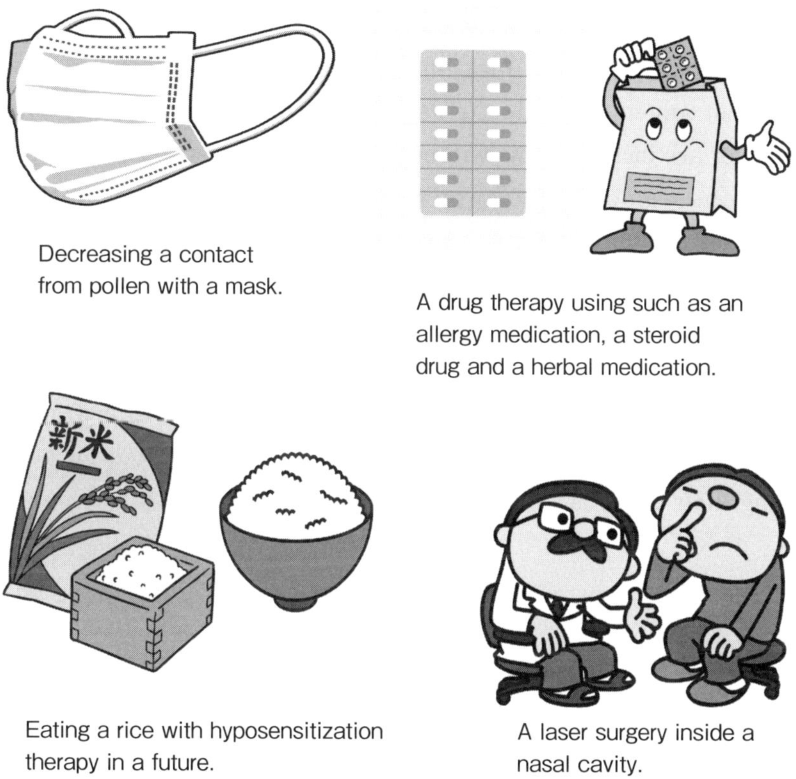
Decreasing a contact
from pollen with a mask.
A drug therapy using such as an
allergy medication, a steroid
drug and a herbal medication.
新米
Eating a rice with hyposensitization
therapy in a future.
A laser surgery inside a
nasal cavity.

Figure 6.4　Methods for decreasing pollen allergy

comprising 10–20 amino acids linked together) is introduced into rice. An experiment conducted with mice showed that the severity of hay fever was relieved by 70 %. Issues such as safety, toxicity, and carcinogenicity, and how to distribute this hay-fever relieving rice commercially, are currently being worked out.

According to a survey conducted by the government, hay fever was ranked second among problems that subjects wanted solved in

---

Note：hyposensitization「減感作：アレルギー症状を起こす原因物質（スギ花粉など）の抽出液等を，数年に渡り少しずつ注射し，体を徐々に慣れさせていく治療法」

the future in the areas of science and technology (an effective technology to prevent metastasis of cancer was first, and Alzheimer's disease was third). In fact, many people suffer from hay fever (foremost among them due to Japanese cedar); if those likely to get hay fever are included, this will be a massive number in the future. An enormous amount of money is also being spent annually on treatments for hay fever. At the end of 2006, 24 million people were said to have hay fever, which is one in five citizens. Gross calculation of treatment costs shows that 193 billion yen has been paid out through healthcare insurance. This amount increases further if over-the-counter medicines are included. The hay fever issue warrants monitoring, not only for those suffering from hay fever, but for people paying their taxes as well.

References

1. Books

1) 斎藤洋三，井手　武，村山貢司：新版 花粉症の科学，化学同人（2006），ISBN 978-4-759-81050-9

2. Journal

1) H. Takagi, et al.：A Rice-Based Edible Vaccine Expressing Multiple T Cell Epitopes Induces Oral Tolerance for Inhibition of Th2-mediated IgE Responses. Proc. Natl. Acad. Sci. USA **102**：pp.17525-17530（2005）

3. URL（2013 年 9 月現在）

1) http://www.nlm.nih.gov/medlineplus/ency/article/000813.htm
2) 花粉の出ないスギ開発進む　来春試験植樹 和歌山県林業試験場 http://www.agara.co.jp/modules/dailynews/article.php?storyid=196683
3) スギ花粉症緩和米の研究開発について　農業生物資源研究所 http://www.nias.affrc.go.jp/gmo/simple.html

---

Note：Alzheimer's disease「アルツハイマー型認知症」

Exercises

1. つぎの 1 ～ 84 の語に対応する英単語または英熟語を本文から選び出し，発音しなさい（動詞は原形を記入しなさい）。

	Japanese	English		Japanese	English
1	圧倒的な		19	吐き気	
2	アレルギー性の		20	腹痛	
3	スギ		21	皮膚炎	
4	花粉		22	不眠症	
5	浮遊する		23	短気	
6	発生（率）		24	食欲	
7	くしゃみ		25	突発	
8	停滞		26	鼻水	
9	咳		27	捕らえる	
10	たん		28	刺激する	
11	ぜんそく		29	誘因	
12	結膜炎		30	越える	
13	かゆみ		31	おたふくかぜ	
14	副鼻腔炎		32	はしか	
15	気管支炎		33	結合する	
16	炎症		34	受容体	
17	胃腸の		35	推測	
18	下痢		36	遺伝する	

	Japanese	English		Japanese	English
37	劣性		58	阻害物質	
38	優性		59	拮抗性	
39	対立遺伝子		60	阻害剤	
40	抵抗性		61	ステロイド	
41	遺伝子		62	抑制する	
42	汚染		63	ヒスタミン	
43	感染		64	改善する	
44	生理的な		65	脱感作法	
45	自律した		66	処置	
46	遺伝的な		67	順化する	
47	療法		68	一環した	
48	花粉症		69	外科手術	
49	ブタクサ		70	鼻腔	
50	モチクサ		71	傾向	
51	吸入する		72	絡み合わせる	
52	吸入		73	あっけにとられること	
53	循環		74	雄性不稔	
54	気象の		75	増殖する	
55	予測		76	細胞培養	
56	檜		77	食用可能な	
57	媒介者		78	実施	

	Japanese	English		Japanese	English
79	主食		82	毒性	
80	和らげる		83	発癌性	
81	ひどさ		84	転移	

2. つぎの各文が本文の内容と一致するものにはT(True)，一致しないものにはF(False）を，文末の（　　）に記入しなさい。

（1）One study has estimated that the incidence of the pollen allergy is more than half of the Japanese population.（　　）

（2）Hay fever could cause headaches, gastrointestinal symptoms, such as diarrhea, nausea, and stomachache, and dermatitis caused by pollen.（　　）

（3）The pollen from Japanese cedars that bloom can travel less than 100 km.（　　）

（4）Our noses start to itch when we inhale Japanese cedar pollen because of the antigen-antibody reaction we possess as humans.（　　）

（5）Environmental factors are not important for the development of hay fever.（　　）

3. つぎの日本語の各文を（　　）の中の語を用いて英語の文にしなさい（必要があれば動詞を適切な形に変換しなさい）。

（1）抗体抗原反応は，人体を守るためにできた歓迎すべき機能である。(antigen-antibody, function, design)

（2）気道に鼻水が入り炎症が生じると，気管支炎を起こす場合がある。(develop, enter, airway, cause)

（３）環境要因としては，アレルゲンへの露出，大気汚染，感染，生活環境や食習慣などがある。（environmental factor, include）

（４）症状が緩和しないなら，花粉症の治療薬を投与する必要ある。（symptom, subside, administer）

（５）この製品は，脱感作法を利用したものである。（utilize, desensitization）

4. つぎの各問いに英語で答えなさい。

（１）Why does hay fever develop?

（２）What therapies are available when should you happen to develop hay fever?

（３）Can we eat transgenic rice containing allergy remedy?

（４）What would be the biggest medical concern in Japan?

（５）How you can protect yourself from the hay fever?

Chapter 7
Will Biomass Save the World?

In recent years, the terms "biofuel" and "biomass" have become familiar. However, these terms themselves are not new, but have been around for a long time. "Biofuel" is translated as "fuel produced from living organisms" and "biomass" is translated in accord with its etymology of "living organism" (Bio) added to "quantity" (mass). Note that biofuels do not include fossil fuels such as petroleum, coal, and natural gas, which were originally plants that became fossilized.

Global warming is the major reason why biomass has become a topic of interest. Mankind has received many blessings by burning fossil fuels. Owing to the increasing convenience and comfort of life following the industrial revolution, the population exploded. But behind the scenes, Earth's temperature has steadily risen. In particular, the increase in temperature recorded in the last 10 years is as much as in the 100 years before that. Everyone living on Earth must be aware of the scorching summer heat and short winters. It is as if we were in a greenhouse, because the carbon dioxide generated by the burning of fossil fuels has a high heat-retaining effect (greenhouse effect). This is global warming. If we continue to use fossil fuels as we have until now, Earth's temperature will rise ever more rapidly, and this will be accompanied by various effects. Actually, all of us are already feeling these effects, and they will appear in more severe forms in the future. The use of biomass instead of fossil fuels has come under focus as a way to control the rate of global warming.

Biofuels that can replace fossil fuels are garnering the most

---

Notes：etyomology「語源，語源学」，greenhouse effect「温室効果」

attention. The raw plant material that forms the basis of such biofuels ranges widely from sugarcane or corn, to trees, seaweed, and sewage sludge as well as other waste. Currently, bioethanol from sugarcane is the most commonly produced biofuel. In fact, the history of this biofuel is surprisingly long. In order to protect the sugar industry from large price fluctuations, Brazil established a law in 1931 to convert part of its sugarcane crop into ethanol for automobile fuel whenever the price of sugarcane was low. This bioethanol is mixed with normal gasoline; nowadays, E22–25 and E100 fuels, consisting of 22–25 % and 100 % ethanol, respectively, are used to power all of the country's automobiles. Furthermore, bioethanol is now an important foreign currency resource for Brazil, which exports the fuel overseas as well.

The method by which ethanol is extracted from sugarcane involves the use of "bagasse" — the pomace left over after crushing the stalks — as a raw material (**Figure 7.1**). The molasses, or waste liquid remaining after sugar production, is mixed with whole squeezed cane juice and fermented; the bagasse is then burned to generate thermal energy; and the fermented ethanol liquor is distilled. The result is what we call bioethanol. The proportion of molasses and the whole squeezed juice fluctuates depending on the international price of sugar. Bagasse is also used for power generation. The energy output–input ratio which is the energy returned on energy invested (EROEI or ERoEI) [the ratio of the amount of usable energy acquired from a particular energy resource to the amount of energy expended to obtain that energy resource] for bioethanol based on sugar cane as the raw material is higher than, for example, that for bioethanol based on corn. The energy output–input ratio indicates the proportion of energy obtained to the amount of fossil fuels used in its production. In the case of sugar cane, the energy output–input ratio is 6, which is to say that 1/6 (17 %) is from fossil fuels and the remaining 83 %

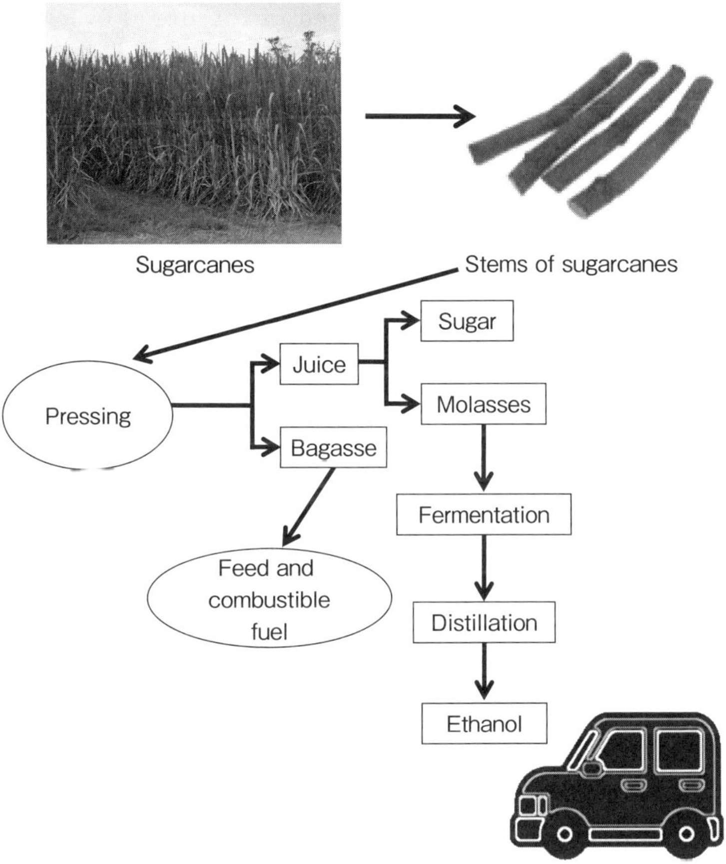

Figure 7.1 The process to produce bioethanol from sugarcane.

contributes to the prevention of global warming. This is called the contribution to global warming prevention (or degree of greenhouse gas reduction). In contrast, the energy output–input ratio for corn is 1.3, with 77 % from fossil fuels and only a 23 % contribution to global warming prevention. The primary reason for this is that the bagasse is used in the production of ethanol from sugar cane.

Of the petroleum consumed in Japan, 20 % is used as raw materials for petrochemical plants to produce plastics and synthetic fibers, which are intimately involved in the life of every person. Currently, biomass is used as raw material in the development of various products. The bioplastic for which commercialization is most advanced is polylactic acid (PLA), a polymer produced by polymerizing monomers of lactic acid. Biomass is first saccharified, after which it is fermented with lactic acid bacteria. The lactic acid thus produced is then polymerized to become PLA. The plant-based PLA is biodegradable; that is, the PLA is decomposed by microorganisms in the soil, ultimately breaking down to water and carbon dioxide. The rate of decomposition is said to be around 6 months to around 3 years; if PLA is buried in earth, it becomes water and carbon dioxide, and then disappears. Many other biodegradable plastics have been developed. A variety of strength, plasticity, and biodegradability can be conferred upon petroleum-based plastics by mixing them with these biodegradable plastics. These plastics are also known as green plastics (**Figure 7.2**).

The problem with biodegradable plastics is that their production requires a large amount of energy. If they are disposed of quickly after being used only once, they are converted to carbon dioxide without the use of energy. Therefore, methods to convert the polymer back to monomers and recycle them by re-polymerization are being considered. When biodegradable plastic is used as a product, carbon dioxide is contained within it. Therefore, returning the plastic quickly to the earth or burning it is considered the last resort, whereas repeatedly recycling the plastic in a different form turns out to contribute greatly to the prevention of global warming.

Biomass is attractive because it is renewable and sustainable. However, if we ignore this aspect and continue to cut down trees

---

Notes：polylactic acid (PLA)「ポリ乳酸」, saccharified「単糖への分解」

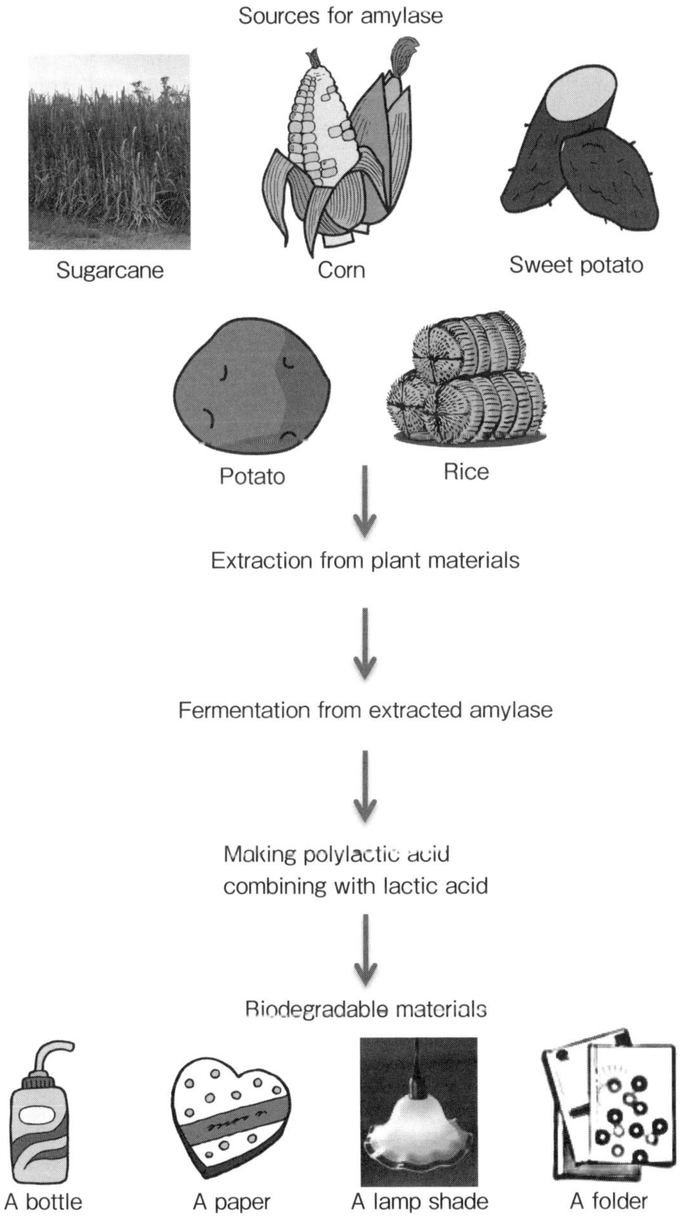

Figure 7.2 The process to create biodegradable products.

rapidly, to obtain bioenergy for example, the area of forests will of course decrease, and the principle of carbon neutrality will collapse. If this is repeated, mountains will increasingly denuded, renewal will become impossible, and the earth's temperature will further increase. This, in turn, will cause the ecosystem to fall out of balance, increase the number of organisms becoming extinct, and at the same time lead to the prolific emergence of pests and weeds. It is said that 100 species become extinct every day. This rate will increasingly accelerate if the warming trend continues its acceleration.

"Sustainable development" is the most important keyword for the development of the human species. And conservation and uses of the biomass, are the parts of the crucial pillars of countermeasures against global warming to support the sustainblity. Monitoring by itself does not maintain anything. "Monitoring becomes necessary to ensure that this continuity is maintained; monitoring ranges from that on a small scale to so-called remote sensing on a large scale involving the use of satellites and aerial photographs. This monitoring information is used to assess deforestation and forest fires. In addition, by using the information from ground measurements used to confirm aerial search results (ground truth) together with information on crop yields and pest occurrences, valuable information can be obtained. Using this information to maintain ecosystems is the duty of all of mankind today so that not only our generation but generations further down the road can utilize them and reap their blessings.

References

1. Books

1) 木谷　収：バイオマスは地球環境を救えるか，岩波書店（2007），ISBN978-4-

Note：aerial photograph「航空写真」

005-00578-9
2) 小泉達治：バイオエタノールと世界の食料需給．筑波書房（2007），ISBN978-4-811-90320-0
3) 坂内　久：燃料か食料か—バイオエタノールの真実，日本経済評論社（2008），ISBN978-4-818-82020-3
4) 川島博之：世界の食料生産とバイオマスエネルギー— 2050年の展望，東京大学出版会（2008），ISBN978-4-130-72102-8
5) 横山伸也，芋生憲司：バイオマスエネルギー，森北出版（2009），ISBN978-4-627-94721-4
6) 産業技術総合研究所：きちんとわかる木質バイオマス（産総研ブックス），白日社（2009），ISBN978-4-891-73125-0
7) 生分解性プラスチック研究会 編：トコトンやさしい生分解性プラスチックの本（B&Tブックス—今日からモノ知りシリーズ），日刊工業新聞社（2004），ISBN 978-4-526-05330-6
8) 日本バイオプラスチック協会 編：バイオプラスチック材料のすべて，日刊工業新聞社（2008），ISBN978-4-526-06122-6

2. Journal

1) http://www.journals.elsevier.com/biomass-and-bioenergy/
2) http://uscip.org/JournalsDetail.aspx?journalID=32
3) http://www.springer.com/engineering/energy+technology/journal/13399

（この他にも多数の研究誌あり）

3. URL（2013年9月現在）

1) http://www.chem-station.com/yukitopics/baiomas.htm（バイオプラスチック）
2) http://www.erec.org/fileadmin/erec_docs/Projcet_Documents/RESTMAC/Brochure5_Bioethanol_low_res.pdf（バイオエタノール）
3) http://www.biodiesel.org（バイオディーゼル）

Exercises

1. つぎの1～59の語に対応する英単語または英熟語を本文から選び出し，発音しなさい（動詞は原形を記入しなさい）。

	Japanese	English		Japanese	English
1	化石		2	化石化する	

	Japanese	English		Japanese	English
3	石油		24	廃糖蜜	
4	石炭		25	残存する	
5	地球温暖化		26	絞る	
6	人類		27	発酵	
7	祝福する		28	産出	
8	燃焼する		29	投資	
9	革命		30	貢献	
10	人口		31	予防	
11	爆発する		32	減少	
12	渇く		33	消費する	
13	二酸化炭素		34	石油化学の	
14	同伴する		35	合成の	
15	集める		36	商業化	
16	海藻		37	単体	
17	下水汚泥		38	重合体	
18	変動		39	重合する	
19	車両		40	生分解性の	
20	外貨		41	分解する	
21	茎		42	微生物	
22	搾りかす		43	消失する	
23	粉砕する		44	強さ	

	Japanese	English		Japanese	English
45	柔軟性		53	生態系	
46	与える		54	出現	
47	処理する		55	増殖する	
48	変換する		56	加速	
49	更新		57	決定的な	
50	持続的な		58	森林減少	
51	原則		59	刈り取る	
52	崩壊（する）				

2. つぎの各文が本文の内容と一致するものにはT(True)，一致しないものにはF(False）を，文末の（　　）に記入しなさい。

（1）“Biofuel” is translated as “fuel produced from fossils”（　　）

（2）Biomass is the new term made recently.（　　）

（3）Biofuel cannot be made from sewage sledge.（　　）

（4）The history of this biofuel is surprisingly short.（　　）

（5）Bioethanol is now an important foreign currency resource for Brazil.（　　）

3. つぎの日本語の各文を（　　）の中の語を用いて英語の文にしなさい（必要があれば動詞を適切な形に変換しなさい)。

（1）バイオマスの語源は生物（Bio）にマス（Mass）つまり量を加えたもので生物量と訳される。(translate, ethymology)

（2）石油，石炭や天然ガスなど植物が化石化してできた化石燃料はバイオ燃料としない。（Biofuel, fossilize, include）

（3）PLA の分解程度は，6 カ月から 3 年くらいといわれている。（poly-lactic acid（PLA）, decomposition, say）

（4）サトウキビ由来のバイオエタノールは，もっとも一般的に作られているバイオ燃料である。（Produce, sugarcane, popularly）

（5）生分解性のプラスチックの問題点は，大量のエネルギーを生産に必要とすることである。（biodegradable, require, production）

4. つぎの各問いに英語で答えなさい。

（1）If we continue to use fossil fuels as we have until now, what will happen?

（2）Do we have any alternatives instead burning to get energies?

(3) Why is the biomass is attractive globally?

(4) How can you make the contribution to global warming prevention?

(5) Can we proceed sustainable development?

Discuss Corner 1

Transgenic Crops Can Help the World ?

The world community faces daunting challenges. Over one billion people are malnourished, often resulting in chronic diseases and premature deaths. Agriculture impacts the environment through pesticides, fertilizers, irrigation, ploughing and conversion of natural habitats. The situation is compounded further by the growth of the world population and climate change.

(EU 2012. EU GMO POLICIES, SUSTAINABLE, FARMING AND PUBLIC, RESEARCH. http://greenbiotech.eu)

Discuss if a technology such as transgenic crops made by genetic engineering can mitigate the present food security concern in the world?

Chapter 8
An Uninvited Guest：Echizen Kurage（Nomura's Jellyfish）

When you hear the word "jellyfish", many of you may recollect being stung on the leg while swimming at the beach. In recent years, the bioluminescent jellyfish *Aequorea victoria* has been in the limelight. The substance in this jellyfish that emits fluorescent light is called Green Florescent Protein（GFP）, and it has become an indispensable research tool in the world of biology. GFP becomes visible when inserted into genetically modified organisms, and by using GFP, it has become possible for researchers to identify individuals and accurately capture changes and movements. It is an indispensable presence in the fields of molecular biology and medicine now. The Nobel Prize in Chemistry 2008 was awarded jointly to Osamu Shimomura, Martin Chalfie and Roger Y. Tsien "*for the discovery and development of the green fluorescent protein, GFP*".（http://www.nobelprize.org/nobel-prizes/chemistry/laureates/2008/index.html）.

There is another jellyfish that has recently been creating an even greater sensation than *Aequorea victoria*. That is the Echizen kurage or Nomura's jellyfish（*Nemopilema nomurai*）, which achieves immense proportions when mature. Previously, populations of Nomura's jellyfish were thought to increase explosively about once every 40 years. However, the interval between harmful outbreaks has shortened since the 1990s, and outbreaks have been reported nearly every year for the past few years. Apart from Nomura's jellyfish, moon

---

Notes：genetically modified organisms「遺伝子組換え生物（transgenic organisms）」
moon jellies（*Aurelia*）「ミズクラゲ」

jellies (*Aurelia*) also cause severe harm. How is it that a creature that has existed on Earth far longer than humans has become such a problematic organism for humans? (**Figure 8.1**)

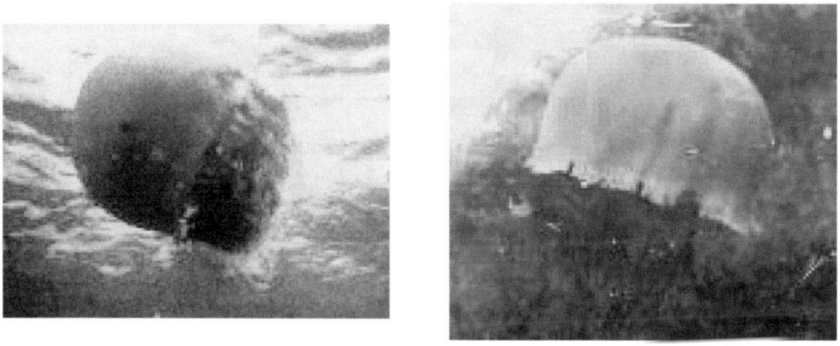

Figure 8.1 The enlarged Nomura's jellyfish (*Nemopilema nomurai*) grown to 150 kg to 200 kg.

First, let us learn about the life cycle of jellyfishes (**Figure 8.2**). The cycle can broadly be divided into the 1) polyp (asexual generation) and 2) jellyfish (sexual generation) stages. Taking the moon jelly as an example, a mature jellyfish first lays eggs. When the fertilized eggs hatch, jellyfish larvae (asexual generation) called planulae are born. These planulae separate from the parent body (that is, the mature jellyfish) and begin to swim into the sea while moving in a spiral. Eventually, the planulae becomes polyps and adhere themselves to seawalls, seaweed, or seashells; at that time, they lose their cilia and become unable to move for some time. They then begin to develop a mouth. Eventually, protrusions at the edges of the mouth and nematocysts that produce venom are formed. The polyps continue to grow and separate one by one as they increase in size (called strobila). At this point, they become immature medusoid (free-living) jellyfish that are autonomous. With further growth, sexual identification becomes possible. The males and females

Note : asexual generation「無性世代」

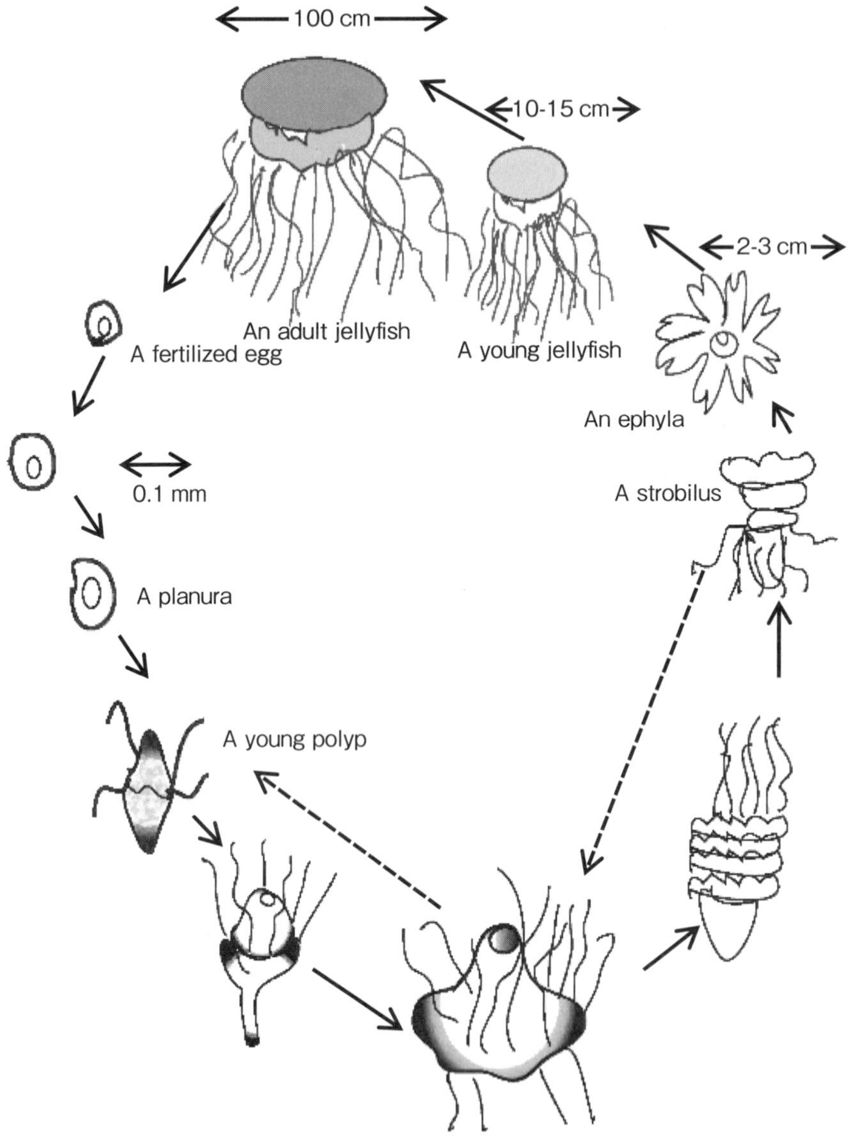

Figure 8.2 A life cycle of Nomura's jellyfish, modified from Nomura's jellyfish and Aurelia.

reproduce and grow, and the next generation of polyps is born from the mature jellyfish. The 1) polyp and 2) jellyfish stages are repeated in the life cycle of jellyfish.

Why then have Nomura's jellyfish and moon jellyfish come to cause such huge problems? The reason is deeply related to the ocean environment. First, it has become increasingly likely that the sites from which Nomura's jellyfish originate are 1) the southwest and southeast coasts of the Korean Peninsula; 2) the East China Sea, Shanghai coast, Yangtze River estuary, and central to northern coastal area of the Yellow Sea; or 3) Tsushima Island as well as a portion of the northwestern coast of Kyushu. A subsequent survey reduced the likelihood that the Japan Sea coast was the source of the outbreaks because the salt concentration is high and the polyp growth rates are poor. In all other regions, there has been rapid construction of embankments and piers rapid construction of embankments and piers and revetment developments associated with economic growth. This has secured places to which polyps could potentially adhere.

However, this alone cannot explain the massive outbreaks of large jellyfish. Another factor is the eutrophication of seawater. Disposal of domestic wastewater causes harmful algal blooms, namely so-called red tides. Red tides consist of zooplankton, which is an ideal food for the polyps.

There is one other important factor, which is the warming of seawater. Warm seawater is advantageous to the growth of polyps. Small Nomura's jellyfish ride the currents across the Japan Sea; some of these continue northward and manage to reach Hokkaido. Nearly all of the remaining jellyfish pass the Tsugaru Strait and reach the Pacific Ocean. Nomura's jellyfish were observed off the coast of Ibaraki prefecture in 2005 and off the coast of Wakayama prefecture in 2009.

By the time they arrive, they are more than 1 000 times the size they were when they started their journey. And as they continue to travel, the jellyfish continue to get more and more enormous. The optimal temperature for Nomura's jellyfish is between 15 ℃ and 29 ℃, and the reproductive period is thought to be between August and December. The miniscule larvae prey on plankton and begin undergoing metamorphosis 1–2 weeks later. By about 1 month, they become young jellyfish more than 10–50 cm in size. They become enormous as they continue their transit, and reach a maximum of 2 meters. This amounts to an astounding 150–200 kg per jellyfish.

Jellyfish can also survive at low water temperatures. Eggs collected during a period of warmth were still fertilizable after they were stored for 3 days not only at room temperature but even in a refrigerator at 2 ℃. Given that they are fully active even after low-temperature storage, they could probably adequately survive even the winter period of the Japan Sea. The jellyfish might also adapt gradually to low water temperatures.

In addition, Nomura's jellyfish spawns massive numbers of eggs several times a year. A single spawning consists of an average of about 3 million to 420 million eggs. Considering that they produce this many eggs and that several jellyfish are thought to grow from a single polyp, it is understandable that the coastlines of the Japan Sea and Pacific Ocean are completely filled with jellyfish. The factors underlying the explosive proliferation of Nomura's jellyfish are 1) coastal development, 2) blooms of zooplankton due to water contamination, and 3) rising sea water temperatures due to global warming, each of which is intimately related to human activity. Here, the greatest fear of those involved is whether Nomura's jellyfish, like the moon jelly, might overwinter, spawn, and breed along the Pacific Ocean coastline. There are numerous sites along the Pacific Ocean coastline that are hospitable to the growth of polyps.

Now, the damage done by Nomura's jellyfish is becoming more and more severe by the year. Moreover, in the past few years, damage has been reported nearly every year. The moon jellyfish causes damage as great as Nomura's jellyfish. By way of example, when there was a massive outbreak of moon jellyfish in Tokyo Harbor in the 1980s the following problems occurred:

1) When they get caught in fishing nets, it takes more time to haul them in.
2) The mesh gets plugged or changes shape, and the catch declines accordingly.
3) Sorting the catch becomes difficult and time consuming.
4) The body temperatures of the fish rise by 3–4 ℃ owing to the mucosal fluid of the jellyfish, and freshness of the catch thus declines.
5) Owing to the nematocyst venom of the jellyfish, the vigor and freshness of the fish decline. The catch will either tumble in price or be unsellable.

In addition, when the number of jellyfish is very high, hauling in the net is impossible and there is no alternative but to cut the net. Naturally, fishing is temporarily put on hold, and the economic burden of those involved in fishing becomes immeasurable. Jellyfish cause fishing damages of around 10 billion yen a year.

Other problems involving jellyfish include the following:

6) Jellyfish occasionally invade fish preserves of yellowtails (*Seriola quinqueradiata*) or red sea breams (*Pagrus major*). At times they also cause mass mortality (death by debilitation).
7) In some cases, massive outbreaks of jellyfish cause mass mortality of shorebirds as well as withering of seaweeds such

---

Note : red sea breams (*Pagrus major*)「マダイ」

as brown macroalgae (*Sargassum muticum*) and eelgrass.

8) At beaches, jellyfish may sting people and lead to prohibition of swimming.

9) In addition, thermal and nuclear power stations have also suffered damages from jellyfish. The water intake for power generation is pumped from deep layers at least 10 m below sea level. During this process, jellyfish are occasionally pumped up with the seawater (in the case of moon jellyfish).

Comparable or greater damage has been reported in the case of massive outbreaks of Nomura's jellyfish.

By what means, then, can we prevent damage from jellyfish? The first is to prevent the eutrophication of seawater, which is the root cause of the massive outbreaks of jellyfish. For this, an international system of cooperation is indispensable. Similarly, embankment developments need to be changed to structures that prevent the adhesion of polyps to the greatest possible extent. Although international coordination is essential for these two steps, this will be a very difficult issue to resolve.

Next, researchers and those involved in the fishing industry are trying to construct new nets by devising ingenious meshes. They are contemplating methods by which to capture only the fish while releasing the large jellyfish from above. In addition, methods to finely shred the captured jellyfish with stainless wire or piano wire are also being considered. Although this method itself is effective, the sea floor may become contaminated if the shredded jellyfish is disposed of in the sea. Therefore, those involved are going through a process of trial-and-error to see whether they can effectively use the shredded jellyfish in a different manner without disposing of it in the sea.

Note : macroalgae (*Sargassum muticum*)「大型藻類」

Another fact is that the harvests of filefish (*Stephanolepis spp.*) and parrot bass (*Oplegnathus fasciatus*), which prey on jellyfish, are declining dramatically because of overfishing. The threadsail filefish (*Stephanolepis cirrhifer*) and black scraper (*Thamnaconus modestus*) prey on an amount of jellyfish up to 10–30 times their own weight. If these predators decline, the numbers of jellyfish will naturally increase. In 2010, along with the massive outbreak of Nomura's jellyfish, the catch of black scraper was bountiful. However, the price fell because so many were caught, and unfortunately for those involved in fishing, the bumper crop did not result in any profits. Also, harvests of Japanese sardines and pilchards, which prey on the same plankton as does Nomura's jellyfish, are undergoing sharp fluctuations because of predation and competition for prey. Methods to prevent overfishing must also be considered. This requires not only the cooperation of individual fishermen, their regions, and country, but also international agreements mainly involving organizations such as the United Nations.

At thermal and nuclear power stations, seawater is used as cooling water during power generation. However, the intake of jellyfish through the water intake ports has become a major issue. When a jellyfish is sucked in, the friction generated causes malfunctions in the electrical system and stops the generator. Efforts are being made to exploit the nature of jellyfish to prevent them from approaching the water intake ports. As jellyfish are affected by sound pressure, brightness, wavelength of light, and electrical current, the

---

Notes : filefish (*Stephanolepis spp.*)「カワハギの類」
parrot bass (*Oplegnathus fasciatus*)「イシダイ」
threadsail filefish (*Stephanolepis cirrhifer*)「カワハギ」
black scraper (*Thamnaconus modestus*)「ウマヅラハギ」

pulsation of their umbrellas can be accelerated or suppressed. By using an intense blinking light or a low-illumination light, the jellyfish can be distanced from the water intake port and made to aggregate near the sea surface. However, it is also true that capital investment is a bottleneck.

Attempts to make effective use of the captured jellyfish are also being carried out. Firstly is to eat them. Among the jellyfish currently prepared and eaten as cold dishes, it is Bizen jellyfish that are primarily used. With the intent of using the bountiful catches of Nomura's and moon jellyfishes, processing methods have been elaborated so that food products with nearly the same texture as Bizen jellyfish can now be made.

The nearly calorie-free jellyfish have also been dried and used as an ingredient in cookies. The low-calorie cookies have become a best-selling product, much of which is consumed in the Kansai region.

New attempts to utilize jellyfish involve the moon jellyfish, for example, which contain collagen, docosahexaenoic acid (DHA: useful for preventing lifestyle diseases), and eicosapentaenoic acid (EPA) in the same proportions as in the Pacific saury (*Cololabis saira*). A functional substance for blood pressure control has also been discovered in the mature jellyfish. These jellyfish can be used to produce large amounts of dietary supplements. In addition, a method to purify lectin, a type of glycoprotein used to diagnose and treat cancer and leukemia — from Nomura's jellyfish has been established. Alternatively, as giant jellyfish contain antibacterials and large amounts of mucin, a glycoprotein with high moisturizing effects, research on effectively utilizing these components has also begun. Dried jellyfish

---

Notes : docosahexaenoic acid (DHA)「ドコサヘキサエン酸」
eicosapentaenoic acid (EPA)「エイコサペンタエン酸」
Pacific saury (*Cololabis saira*)「サンマ」

is also being utilized as fertilizer for fields of rice and other crops as well as farmed fish and livestock. Furthermore, the massaging effect of jellyfish pulsation is effective for the treatment of nerve pain and arthritis; it is already utilized in Norway as jellyfish therapy.

As shown above, there are ways to effectively utilize plagues of jellyfish. Nevertheless, a fundamental solution has yet to be found. Unless fundamental efforts toward stemming massive proliferation are conducted on an international level, the conditions are likely to increasingly worsen. Is the day near when the seas of Japan are completely filled with jellyfish?!

References

1. Books

1) 日本水産学会監修，安田　徹著：エチゼンクラゲとミズクラゲ—その招待と対策—，（ベルソーブック）成山堂書店（2007），ISBN978-4-425-85301-4

2. Journal

1) T. Kamiyama : Planktonic Ciliates as a Food Source for the Scyphozoan Aurelia Aurita (s.l.) ; Feeding Activity and Assimilation of the Polyp stage. J. Exp. Mar. Biol. & Ecol. **407**, pp.207-215 (2011)
2) K. Ohtsu, et al. : Experimental Induction of Gonadal Maturation and Spawning in the Giant Jellyfish *Nemopilema nomurai* (Scyphozoa: Rhizostomeae). Mar. Biol. **152**, pp.667-676 (2007)

3. URL (2013 年 9 月現在)

1) http://www.springerimages.com/Images/RSS/5-10.1186_1472-6750-9-98-0

Exercises

1. つぎの 1 ～ 82 の語に対応する英単語または英熟語を本文から選び出し，発音しなさい（動詞は原形を記入しなさい）。

	Japanese	English		Japanese	English
1	招かれざる		19	ポリプ	
2	クラゲ		20	防波堤	
3	回顧する		21	繊毛	
4	刺す		22	刺胞	
5	生物発光性の		23	毒	
6	注目の的		24	自由遊泳有性相	
7	放射する		25	自主的な	
8	不可欠の		26	河口	
9	可視の		27	続いて	
10	捉える		28	築堤	
11	成熟した		29	桟橋	
12	巨大な		30	護岸	
13	感覚，評判		31	大きい	
14	産卵する		32	富栄養化	
15	受精する		33	赤潮	
16	幼生		34	動物プランクトン	
17	プラヌラ（幼生）		35	流行，流れ	
18	螺旋の		36	生殖の	

	Japanese	English		Japanese	English
37	極小		58	捕食する	
38	仰天する		59	潤沢な	
39	産卵		60	豊作	
40	増殖		61	収益	
41	汚染		62	イワシ	
42	緊密に		63	捕食	
43	海岸線		64	吸い込む	
44	手厚い		65	摩擦	
45	塞ぐ		66	故障，不調	
46	たぐる		67	ちらつく	
47	限りない		68	集める	
48	侵略する		69	障害	
49	致死率		70	コラーゲン	
50	しぼむ		71	栄養補助(サプリ)	
51	アマモ		72	精製する	
52	禁止		73	レクチン	
53	考案する		74	糖タンパク	
54	器用な		75	抗細菌性の	
55	熟考する		76	ムチン	
56	裁断		77	湿気を与える	
57	試行錯誤		78	肥料	

	Japanese	English		Japanese	English
79	脈拍		81	大量発生，疫病	
80	関節炎		82	茎，柄	

2. つぎの各文が本文の内容と一致するものにはT(True)，一致しないものにはF(False) を，文末の（　　）に記入しなさい。

（1）GFP is an indispensable presence in the fields of molecular biology and medicine.（　　）

（2）Red tides consist of organic compounds but not planktons.（　　）

（3）Jellyfish can not survive at low water temperatures.（　　）

（4）Jellyfish cause fishing damages of around 10 billion yen a year.（　　）

（5）Jellyfish can be used for food, cosmetics and medicine.（　　）

3. つぎの日本語の各文を（　　）の中の語を用いて英語の文にしなさい（必要があれば動詞を適切な形に変換しなさい)。

（1）エチゼンクラゲは成熟すると巨大になる。(immense, mature)

（2）これだけではなぜ巨大なクラゲの大量発生を説明できない。(explain, massive, outbreak)

（3）一回の産卵で少なくとも300万の卵が生まれる。(spawn, consist of)

（４）魚の乱獲を防ぐ方法を考慮する必要もある。
（overfishing, prevent）

（５）クラゲは食用としてはほぼカロリー皆無であるが，乾燥してクッキーの原料として使える。（calorie-free, ingredient）

4．つぎの各問いに英語で答えなさい。

（１）State the life cycle of the jerry fishes.

（２）Why then have Nomura's jellyfish and moon jellyfish come to cause such huge problems?

（３）What are damages that could be caused by the Nomura's jellyfish?

（４）By what means, then, can we prevent damage from jellyfish?

（５）What would be an effective use of the captured jellyfish?

Chapter 9
And Then There Were No Bees

In the autumn, plants bear many fruits of varying color. We sense the changing seasons by watching these fruits. Various fruits also make the dinner table a blessing; they can make for menus that are nutritionally balanced as well. However, fruits that should be appearing in the fall are beginning to fail. This is happening with almonds. The reason is not because almond trees are weakening or dying. Almond trees bear snow-white flowers in early February in the northern hemisphere. Almond flowers are not "wind-pollinated flowers" whose pollen is carried by wind. They are "insect-pollinated flowers" that enlist the help of insects to transfer the pollen from the stamen to the pistil for cross-fertilization. Of the many insects, the European honeybee is the most capable pollinator. American almond producers borrow honeybees at a considerable price from commercial beekeepers to pollinate their flowers. Almonds are also a high-value food. Owing to research reports claiming that vitamin E and antioxidant substances in almonds are effective in preventing heart disease, they have become ever more popular (**Figure 9**).

However, despite the popularity of almonds, their pollinators, the European honeybees, have started to disappear without warning. This phenomenon has been termed "Colony Collapse Disorder" (CCD) and has become a major problem in America. The European honeybee is used for food production in Japan as well, such as in the greenhouse cultivation of strawberries, tomatoes, and melons. The CCD problem has also spread to Japan, and fruit prices are increasing as a result.

Why do European honeybees suddenly disappear? Various speculations flew about. The "cell phone hypothesis" assumed that cell

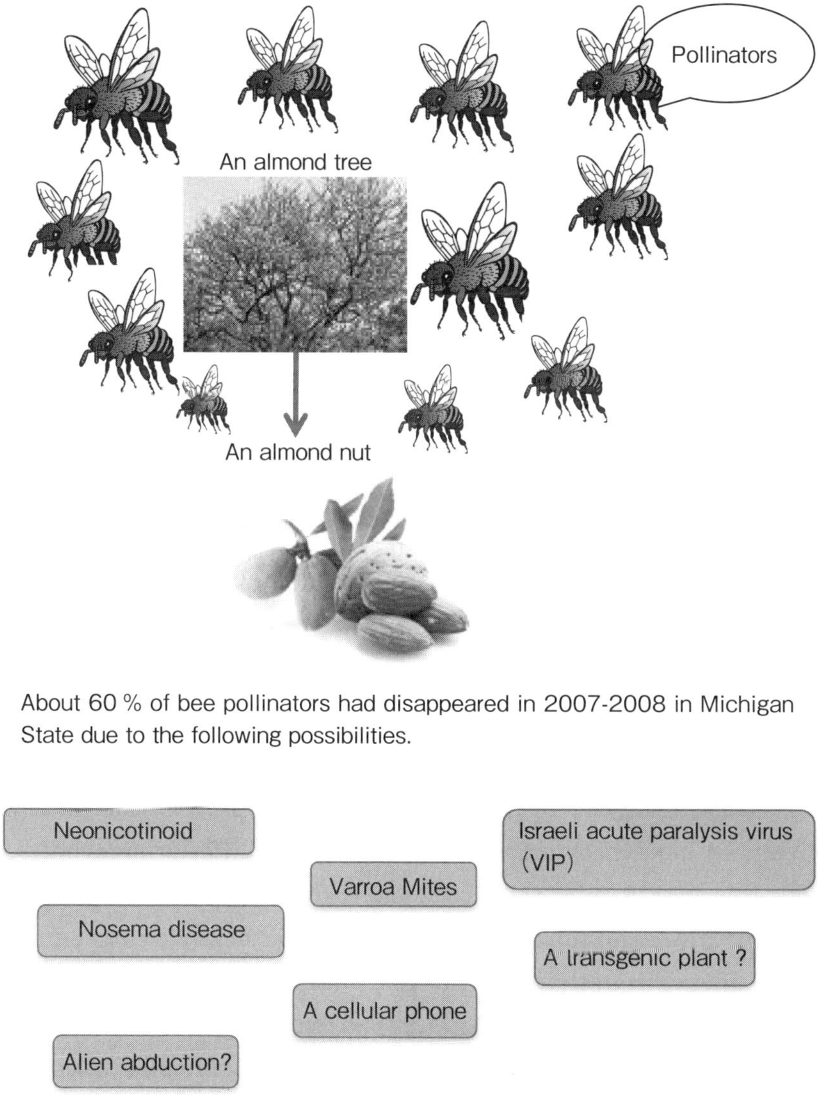

Figure 9 Why did a colony collapse disorder (CCD) occur?

phone radiation affected the bees' sense of direction. Another was "genetically engineered crop hypothesis" where bees were affected by crops that had been genetically engineered to have built-in pest control characteristics. There was even an "alien abduction hypothesis", which claimed that aliens had kidnapped the bees.

Amid these circumstances, the organism that was immediately suspected was the *Varroa* mite. This mite attaches to a honeybee and feeds on its body fluid. Although it attaches to adult bees, it prefers bee larva. An adult mite carrying eggs enters a bee brood cell and attaches itself to a bee larva and gorges itself on the larva's body fluid. It then lays its eggs in the brood cell. The hatched mites again suck the body fluids of the larva and grow, and eventually copulate. They escape outside when the cell is opened, and then move to different cells in the hive and proliferate. Although larvae that have had their body fluids sucked by mites do not die, they become infected by various diseases through the holes opened by the mites, and become emaciated and malnourished. They often become deformed as well. This is why the *Varroa* mite is the organism most feared by beekeepers.

To eradicate this mite, an acaricide called "Apistan" (plastic strips soaked with the insecticide fluvalinate) is used. Given that both mites and honeybees are arthropods, Apistan affects honeybees as well. However, it is of relatively low toxicity, and if the Apistan is removed after the mites are eradicated, the honeybees gradually regain their strength. Nevertheless, mites soon acquire resistance against the agrochemical substance and Apistan becomes completely ineffective. Despite this, some beekeepers still use fluvalinate directly at high concentrations to expel mites. Yet the results are no different, and the mites acquire resistance in no time. It is a cat-and-mouse

Notes：*Varroa*「ダニの属」, brood cell「ハチの巣の穴」

game of insecticides and resistance.

Next to emerge was "Checkmite", an organophosphate insecticide. Although effective in the beginning, the mites soon became resistant again. In the course of investigating this "mite culprit hypothesis", an important fact was revealed. The number of mites in colonies showing no signs of CCD was the same as in those that had become empty because the bees had left. At the very least, it became clear that mites were not the culprits causing CCD.

At around this time, CCD had become a big problem throughout America. The search for the culprit was covered by the mainstream media on a near-daily basis. Prof. W. Ian Lipkin, the discoverer of the "West Nile Virus", took up the challenge of the hunt. Precisely around this time, the entire nucleotide sequence of the European honeybee had just been deciphered. Taking bees from a normal colony, Prof. Lipkin processed them in liquid nitrogen and analyzed the various kinds of genetic information they contained. On the basis of this information, he concluded that the "Israel acute paralysis virus" was behind CCD. This virus, which was discovered in 2004, was found in high probability at colonies affected by CCD. In contrast, it was hardly found at colonies unaffected by CCD. 2004 was also the year in which America began importing honeybees from Australia, as well as the year that CCD was first reported. It appeared that suspicion would rapidly settle on the Israel acute paralysis virus. The U.S. government temporarily halted the importation of bees from Australia. It appeared that this problem had been quickly resolved. However, the actual situation was not so simple. Australian bees are exported not only to America but also to Canada and Europe. However, the CCD problem has not occurred in Canada or Europe. Moreover, the characteristics of Israel acute paralysis virus are that when bees become infected, their wings tremble, paralysis occurs, and then the bees die near the beehive. This was clearly different from the CCD phenomena occurring in America. Who was the culprit?

The next suspect singled out was the microsporidian, *Nosema apis*. However, this microsporidian has been found in frozen samples from 1995 (prior to the occurrence of CCD). Therapeutic drugs are also quite effective against infection by *Nosema apis*. The decisive fact was that bees had disappeared even in colonies where the *Nosema* spore levels were extremely low. There were other cases suspected in this manner. However, as the investigations proceeded, they too were revealed not to be the direct cause of CCD.

Almonds are produced in places around the world, such as Spain, Greece, and Turkey, where conditions are appropriate. However, by far the most dominant producer of almonds is America, which accounts for more than 80 % of total production. America is now experiencing not a gold rush but an "almond rush", with production soaring and demand also rising rapidly owing to the effects of advertising. Things were still good when supply and demand were in balance. However, with the rapid increase in demand, almond producers have increased, resulting in a concomitant and rapid increase in demand for bees. Massive numbers of bees are released into almond orchards to achieve the highest possible cross-pollination rates. Moreover, the time during which almonds can be pollinated is very short — only 3 days after flowering, after which the probability of pollination decreases to zero. There are bound to be various stresses when countless numbers of bees are released for a very short time into wide-open almond orchards.

The most popular almond tree is the variety "Lebanon". Given that almonds are self-incompatible, another variety is needed for cross-pollination. Then we have a single pollinating insect, the

---

Note：*Nosema apis*「単細胞の寄生生物」

European honeybee. To keep the bees healthy, we also have various pesticides for pests. A selected insect, in a selected space, with select insecticides, the situation is as if the bees were just one part of a honey-producing machine. Under such extremely stressful conditions, could the bees have disappeared because they could no longer withstand the duress? No one yet understands the underlying cause. There is but one fact: "And then there were no bees".

References

1. Books

1）石　弘之：地球・環境・人間 II，（岩波科学ライブラリー）岩波書店（2008），ISBN978-4-000-07481-0
2）ローワン・ジェイコブセン：ハチはなぜ大量死したのか（文春文庫），文藝春秋社（2011），ISBN978-4-167-65175-6
3）蜂群崩壊症候群―消えたミツバチの謎，日経サイエンス 2009 年 7 月号，日本経済新聞出版社

2. Journal

1）Cox-Foster DL, et al.：A Metagenomic Survey of Microbes in Honey Bee Colony Collapse Disorder. Science **318**, pp.283-287（2007）

3. URL（2013 年 9 月現在）

1）http://www.ars.usda.gov/News/docs.htm?docid=15572

Exercises

1．つぎの 1 ～ 57 の語に対応する英単語または英熟語を本文から選び出し，発音しなさい（動詞は原形を記入しなさい）。

	Japanese	English		Japanese	English
1	実る		2	栄養のある	

	Japanese	English		Japanese	English
3	半球		24	巣	
4	風媒の		25	ダニ駆除薬	
5	虫媒の		26	撲滅する	
6	雄しべ		27	節足動物	
7	雌しべ		28	回復する	
8	交雑受精		29	駆除する	
9	セイヨウミツバチ		30	獲得する	
10	花粉媒介者		31	いたちごっこ	
11	養蜂業者		32	殺虫剤	
12	授粉する		33	抵抗性	
13	抗酸化剤		34	有機リン	
14	現象		35	犯人	
15	広がる		36	塩基配列	
16	放射		37	解読する	
17	組み込んだ		38	移住地，群落	
18	特性		39	液体窒素	
19	宇宙人		40	分析する	
20	誘拐		41	結論づける	
21	谷間，溝		42	確率	
22	孵化する		43	容疑	
23	交尾する		44	停止する	

	Japanese	English		Japanese	English
45	震える		52	無数の	
46	まひ		53	自家不和合性の	
47	微胞子虫		54	殺虫剤	
48	決定的な		55	ストレスの多い	
49	暴露する		56	我慢する	
50	伴う		57	拘束，強要	
51	果樹園				

2. つぎの各文が本文の内容と一致するものにはT(True)，一致しないものにはF(False）を，文末の（　　）に記入しなさい。

（１）Colony Collapse Disorder, does not spread in Japan（　　）

（２）Mites may be the cause of Colony Collapse Disorder.（　　）

（３）Apistan, an common acaricide, becomes completely ineffective.（　　）

（４）Prof. Lipkin concluded that “Israel acute paralysis virus” was behind CCD.（　　）

（５）The majority of almonds are produced in USA.（　　）

3. つぎの日本語の各文を（　　）の中の語を用いて英語の文にしなさい（必要があれば動詞を適切な形に変換しなさい)。

（１）犯人捜索は，ほぼ毎日のように情報メディアの主流となった。(cover, culprit, mainstream)

（2）アーモンドは世界的に人気があるが，その授粉者であるセイヨウミツバチは，なんの予告もなしに消えつつある。(despite, popularity, disappear)

（3）このダニはミツバチに付着し，体液を吸い取る。(attach, suck, fluid)

（4）セイヨウミツバチの全塩基配列が最近解読されたばかりだ。(nucleotide sequence, decipher)

（5）イスラエル急性麻痺ウイルスに疑惑が集中した。(suspicion, settle on)

4. つぎの各問いに英語で答えなさい。

（1）What is the result of Colony Collapse Disorder (CCD) spread in Japan?

（2）What would be the possibilities of CCD?

(3) Where are almonds produced?

(4) How do you prevent CCD?

(5) Do you believe the alien abduction theory?

Discuss Corner 2

Coexistence with wildlife

As human development progressively encroaches on wildlife habitat, conflicts between wildlife and people increase. Each year, in response to such actual or perceived conflicts, people turn to lethal control efforts to kill "offending" animals. In addition to being inhumane, lethal control efforts are generally doomed to fail, as they don't address the root causes of conflicts or provide long-lasting solutions.

参考 http://www.bornfreeusa.org/a7_coexist.php

Discuss how you can support both human activities and wild life conservation practices.

Chapter 10
The Avian Ark：Save Our Endangered Species!

Various organisms are in danger of dying out. Twelve percent of all endangered species are birds, and their rate of extinction is accelerating. At this rate, it is said that approximately 30 % of all bird species will die out by the end of the 21st century. Although attempts to reduce the extinction rate by 2020 were implemented under Convention on Biological Diversity, they hardly had any impact. On the contrary, the rate of extinction has ironically further accelerated.

Birds rank high within the food chain and are said to be greatly affected by changes in the environment. There are bird species that have teetered on the brink of extinction in the past and those that have actually disappeared as well. The most representative example is that of the crested ibis or Japanese crested ibis（*Nipponia nippon*）（**Figure 10**）.

In the not too distant past, the Japanese crested ibis was commonplace in Japan's paddy fields. It was the bird with which the Japanese were most familiar. However, when hunting became widespread in the Meiji period, their numbers declined rapidly, and they were finally designated as protected birds in 1908. Yet their numbers continued to decrease drastically thereafter, and by 1950, the number of wild individuals had declined to less than 40. Subsequently their numbers continued to decline, and by 1970, wild crested ibises had disappeared entirely; the only surviving crested ibises in Japan ended up being those bred in captivity. One reason for this decline is thought to be that the crested ibises ate food that was contaminated

---

Notes：Convention on Biological Diversity「CBD：生物多様性条約」(http://www.cbd.int/)
Japanese crested ibis（*Nipponia nippon*）「トキ（朱鷺）」

A crested ibis or Japanese crested ibis: *Nipponia nippon*

An oriental stork, *Ciconia boyciana*

A Short-tailed albatross, *Phoebastria albatrus*

A pair of ptarmigan, *Lagopus muta*

An Okinawa rail, *Gallirallus okinawae*

A steller's sea eagle, *Haliaeetus pelagicus*

Figure 10 The endangered or extinct birds in Japan.

by agricultural chemicals. Another major factor is that their breeding grounds were deforested. In the end, the last crested ibis named "Kin (gold)" named with a letter kanji of a protection advocator, died in 2003, and the Japanese crested ibis were extinguished. Although there were attempts to artificially incubate and hatch the crested ibises being bred, all attempts ended in failure.

Nonetheless, the enthusiasm of people who love the crested ibis did not wane. They borrowed a pair from China, which has the only remaining wild population and the pair gave birth in 1999 to the first ever chicks bred in captivity to mature uneventfully. The grandchildren of these birds were also born, and so far, more than 100 crested ibises have been bred in captivity. Attempts to reintroduce them into the wild are being carried out.

Incidentally, the first pair of these crested ibises was from China. Are these crested ibises the same as or different from those of Japan? Stuffed specimens of wild Japanese crested ibises that had been preserved in various regions were useful for addressing this question. When the sequences of the DNA extracted from the stuffed specimens were compared with those from the Chinese crested ibis, only eleven sites among the 16 782 total DNA bases (0.066 %) of the mitochondrial DNA sequence differed. This difference was small enough to consider the two species to be identical. Even more surprising was that the base sequence for a stuffed Japanese crested ibis discovered as a carcass in 1926, was identical to that for the crested ibises that are currently bred in Japan.

As a next attempt, conservationists are trying to return the crested ibises bred at the Reintroduction Station into the wild. The birds are trained in 1) foraging, 2) flight, 3) predator avoidance, and

---

Notes : mitochondrial DNA「ミトコンドリア DNA」
Reintroduction Station「再導入試験場」

4) sociality at the Reintroduction Station. Two of these birds became a pair, and the hatched chicks became fledglings uneventfully. Unexpectedly, they managed to end up being trained in 5) breeding. The ultimate goal is to release the birds into the outdoors by using gigantic cages and videotapes to gradually acclimatize them to the external environment. The day will soon arrive when we see the second and third generations of crested ibises that have settled down somewhere in Japan.

Unfortunately in 2010, a terrible tragedy occurred. A Japanese marten, a small animal, infiltrated the gigantic cage through a small hole during the night, causing the crested ibises to panic and nearly all of them to die. Such holes have now been closed so that small animals cannot get inside. But the immeasurable price paid reminds us anew of the difficulty of reintroducing captive animals into the wild.

The oriental white stork (*Ciconia boyciana*) is a species of bird that died out in Japan much like the Japanese crested ibis. Just a little while ago, the oriental stork was a commonly seen bird, yet like the Japanese crested ibis, it became extinct in no time for reasons similar to the case of the crested ibis. The oriental stork breeds along the Amur River basin in eastern Siberia and overwinters in China, Taiwan, Korea, and Japan. Although its external appearance is similar to that of the crane, the oriental stork nests in pine trees and is completely carnivorous, eating loaches, frogs, and crucian carp. In contrast, cranes nest on the ground (e.g., on grass-covered plains) and are omnivorous.

At one time in Japan, the oriental stork foraged for prey in the shallow waters of rice paddies or waterways and rivers, and especially

---

Notes : oriental white stork (*Ciconia boyciana*)「コウノトリ」

Amur River basin「アムール川流域」, crucian carp「フナ」

preferred marshes, where food is plentiful. However, the oriental stork faced overhunting by rifles in the Meiji period; and with deforestation in the Meiji period, environmental destruction during World War II, and especially land reform as well as river improvement, the marshes serving as feeding grounds gradually declined. But it is the use of agricultural chemicals that is thought to be the most potent factor for its extinction. It was 1971 when the last wild oriental stork in Japan died, and an autopsy found the corpse contaminated by agricultural chemicals, in particular by mercurial substances and PCBs.

Although the wild oriental stork has died out in Japan, artificial breeding programs have been in place since 1956. Yet all of the attempts ended in failure. Half of the eggs were infertile, and the other half were abortive in which the hatchlings prematurely stop growing. Experts think that this was likely due to the effects of agricultural chemicals. In this context, a turning point arrived in 1985 when six hatchlings were sent to Hyogo prefecture from her sister-state in Russia, Khabarovsk Krai (mostly in the basin of the lower Amur River). In 1989, the hatchlings that had grown into adults gave birth to long-awaited chicks. The adults that grew from these chicks have also left a succeeding generation. In 1999, the Hyogo Prefectural Homeland for the Oriental White Stork was completed on 175 hectares of land for the purposes of protected breeding and reintroduction into the wild; it continues to carry out activities toward reintroduction. In 2005, five birds were released from this facility, and in 2007, the first chick in 43 years hatched outdoors in an area of rice paddies and grew up to leave the nest. These attempts are not merely aimed at reestablishing the oriental stork, but also have the greater goal of regenerating natural environments. In rice paddies, various

Notes : land reform「土地改革」, PCB「ポリ塩化ビフェニール」

efforts are under way to lure back the loaches and frogs that constitute the oriental stork's prey. For example, the Japanese brown frog (*Rana japonica*) lays eggs between February and March; yet, for typical farming methods, water is drained during this time. Therefore, attempts are being made to secure egg-laying sites by letting water flow into rice paddies during winter as well. And in Toyooka, the task of "*nakaboshi* (mid-season drainage)" is carried out in mid-June, which is just around the time of egg-laying for leopard frogs (*Rana nigromaculata, or Pelophylax nigromaculatus*) and Japanese tree frogs (*Hyla japonica*). Therefore, tests to delay the mid-season drainage by 1 month are being carried out. This is also turning out to help dragonfly larvae undergo adult emergence eclosion, and the oriental stork's food is about to be secured. Meanwhile, tests to reduce or eliminate agricultural chemicals have also begun. Thus far, the oriental storks that have been released into the wild have unfortunately come back to the biotopes or the Homeland for the Oriental White Stork. Nonetheless, the day we see these patient efforts pay off and oriental storks return to the wild is probably quite near. Incidentally, in the West, the stork is a lucky creature that delivers babies. In 2006, Prince and Princess Akishino participated in a ceremony to release four birds. It was soon afterwards that the Princess's pregnancy was announced.

Until now, we have been discussing the recovery of birds that were once locally extinct. Let us now shift the narrative to the comeback of birds designated as endangered species. One such bird is

---

Notes : Japanese brown frog (*Rana japonica*)「ニホンアカガエル」

(common dark-spotted) leopard frogs (*Rana nigromaculata* あるいは *Pelophylax nigromaculatus*)「トノサマガエル」

Japanese tree frogs (*Hyla japonica*)「ニホンアマガエル」

Biotope「ビオトープ，生物群集の生息空間」

the albatross. Do you know what a decoy is? It is an exact model of a bird made of wood, used originally as a lure for hunting. Albatross decoys have become the saviors of albatrosses. The short-tailed albatross (*Diomedea albatrus*, or *Phoebastria albatrus*) is designated as a class II endangered species by the International Union for Conservation of Nature (IUCN). It is the largest among related albatrosses that inhabit regions north of the equator. An adult bird attains a wingspan of 4.2 m and can weigh as much as 5 or 6 kg. The chicks are covered in black feathers and have pink beaks and bluish-gray legs. At around 4 or 5 years of age, the feathers on the breast and abdominal area become white, and the back becomes white by around 6 or 7 years. And by around 10 years of age, nearly the entire body becomes white, with the head covered in gold-colored feathers. Albatrosses are birds that reproduce once a year. The breeding period arrives around mid-October, and a single egg is laid in November. Both parents raise the chick at the breeding grounds; then, after a period of 2 or 3 weeks of fasting, the chick leaves the nest for the sea.

Until the Edo period, hunting was generally banned. For this reason, domestic wild bird species were protected. However, when the hunting ban was lifted in the Meiji period, albatrosses sharply decreased because of overhunting. Furthermore, transportation to distant sites became possible with the development of the steamboat, and albatrosses came to be increasingly overhunted for their feathers, which until then had not attracted attention. Although the Japanese Government designated it as a protected bird, overhunting continued. In a 1930 survey, the number of birds had decreased to only 300, and by 1949, the bird was reported to have died out. However, in 1951

Notes : short-tailed albatross (*Phoebastria albatrus*)「アホウドリ」

International Union for Conservation of Nature「IUCN : 国際自然保護連合」(http://www.iucn.org)

more than ten were confirmed in the Izu-Torishima and Senkaku Islands, and conservation activities were begun. It was confirmed that up to then, the rate of reproductive success was low, with an average of only 44 % of eggs laid resulting in chicks leaving the nest. The reason for this was that the breeding grounds were lacking in vegetation because of the increased load exerted by the albatrosses themselves (use for nesting materials, trampling, excretion). Hence, the reproductive rate was low. When indigenous plants were transplanted on the islands to increase the area of vegetated land, the reproductive rate increased to 67 %. However, the sloping ground and volcanic sand of the islands resulted in landslides and the reproductive success rate again decreased. From previous observations and prevention activities, a consensus was reached that the nests of albatrosses are on steep slopes on remote islands, and that more stable sites should be sought for nest-building.

The present breeding grounds are on the southern slopes of Izu-Torishima and on the Senkaku Islands. First, on Torishima, the breeding grounds were arranged to be translocated to the relatively gentle western slope. Here, decoys and vocalizations of the albatross played important roles. Researchers set up a solar-powered audio station based on decoys at this site, upon which five albatrosses soon alighted. Subsequently, their numbers grew yearly; by 1996, one chick had hatched and safely left its nest. By 2009, 60 pairs and 300-350 birds were estimated to live at this site. The total number for all of Torishima has recovered to 3 000.

Although the breeding activity on Torishima is on the right track, a major problem is that this island is always volcanically active. It is said that half of the albatrosses would be affected were a large earthquake to occur. So a suggestion to create safer breeding grounds was raised, and after various surveys, Mukojima Island, located to the north of the Ogasawara Islands, was chosen as the third breeding ground. From past data, the threat of an eruption there is considered to be virtually non-existent. In addition, this is also the breeding ground for the black-footed albatross (*Diomedea nigripes*, or *Phoebastria nigripes*) and Laysan albatross (*Diomedea immutabilis*, or *Phoebastria immutabilis*), which are species closely related to the short-tailed albatross. The short-tailed albatross is thought to be more likely to breed near identical or closely related species. Moreover, because this island is uninhabited, there is little risk of human interference. There was also the advantage that carrying out surveys and conservation activities would be relatively easy because Mukojima is only a short hop from Chichijima, the center island of the Ogasawara Islands.

The basic method of using decoys and vocalizations was the same; however, given that the geographical conditions on Mukojima differed from those on the Torishima and the Senkaku Islands, and that the techniques for artificially raising short-tailed albatrosses were hardly known, breeding experiments were first conducted in Hawaii (in 2006) with the Laysan albatross (rather than with rare species), for which data were abundant. The results showed that four out of ten birds left their nests. Autopsies on the other birds revealed a high likelihood of food poisoning because the food provided was infected with bacteria. Thus, the cause of death was inadequate food sanitation. Reflecting upon and taking advantage of these observations, researchers attempted to artificially raise the black-footed albatross. In 2007, ten 50-day-old black-footed albatrosses were transported to Mukojima (http://www.acap.aq/latest-news/translocated-short-tailed-and-black-footed-albatrosses-continue-to-do-well-on-japans-mukojima-island); extra precautions were taken regarding food palatability, taking care to prevent indigestion by using a food processor. As a result, although one bird died of indigestion, the remaining nine safely left their nests. Finally, it was the short-tailed albatross's turn. In 2008, ten 40-day old short-tailed albatrosses were translocated from Torishima to Mukojima via helicopter. This time sanitation was adequately controlled and the animals were raised while paying close attention to weight gain. The result was that not a single bird died, and every bird left its nest uneventfully. In addition, fortunately the birds did not become imprinted on people, and they left their nests without becoming accustomed to the breeding staff. This transfer plan is currently ongoing. Short-tailed albatrosses may return in search of partners as early as 2011. If this effort goes smoothly, the species may be removed from the endangered species list by 2045 at the earliest.

Many other bird species are also on the brink of extinction. The Okinawa rail (*Gallirallus okinawae*) is a flightless bird that lives on the main island of Okinawa. Although it has even been designated as a protected species, the number of surviving animals as well as their range has been declining in recent years. The most important factors for this decline have been the mongoose introduced by humans as a predator of snakes and housecats that have become feral. If we examine the range of mongooses, we find hardly any Okinawa rails within it. Mongooses and feral cats are clearly preying upon the Okinawa rail. Hence, efforts to capture mongooses and defend the protected area with netting have begun to try and stop extinction.

There are also bird species whose numbers have dramatically declined because of other causes. The rock ptarmigan (*Lagopus muta*), which is a bird that lives in highland areas, has long been

---

Notes : Okinawa rail (*Gallirallus okinawae*)「ヤンバルクイナ」

mongoose (例えば *Helogale parvula*)「マングース」

rock ptarmigan (*Lagopus muta*)「ライチョウ (grouse)」

exalted by Japanese people as a miracle-working bird. However, global warming has increasingly narrowed their range. And in concert with global warming, the Japanese deer (*Cervus nippon*) and Japanese monkey (*Macaca fuscata*), which did not exist previously in the alpine regions, have extended their ranges and are devouring the alpine plants that constitute food for the ptarmigan. As counter-measures, we must carry out breeding and artificial reproduction programs to increase the number of individuals, as well as continue to protect the ptarmigan by using sanctuaries. As for the current state of global warming itself, we have no other choice but to keep an eye on the issues.

Steller's sea eagle (*Haliaeetus pelagicus*) is also a bird species undergoing drastic change as it is indirectly affected by human actions. One reason is death by lead poisoning from the lead pellets used during hunting. Steller's sea eagle is a carnivore. It suffers lead poisoning by eating the lead in the carcasses of culled Hokkaido sika deer (*Cervus nippon yesoensis*), which have become destructive animals in recent years. Although the use of lead pellets is banned in Hokkaido, thoroughly enforcing this ban is very difficult. Furthermore, many wind turbines have begun operation in Hokkaido in recent years as wind power generation has gained popularity as earth-friendly energy. However, many of these sites have turned out to be habitats of migratory birds such as Steller's sea eagles; alternatively, harmful bird strikes are occurring because the sites are near the eagles' feeding grounds. Apart from these developments, the Sakhalin Project, an oil and gas development on Sakhalin Island in which

---

Notes : Japanese deer (*Cervus nippon*)「ニホンジカ」
Japanese monkey (*Macaca fuscata*)「ニホンザル」
Steller's sea eagle (*Haliaeetus pelagicus*)「オオワシ」
Hokkaido sika deer (*Cervus nippon yesoensis*)「エゾシカ」, Sakhalin「樺太（サハリン）」

Japan is a participant, has also started. Sakhalin is a major breeding ground for the Steller's sea eagle. This breeding ground would receive a major blow were petroleum to leak from pipelines during its transport.

In this way, numerous endangered species are facing a further crisis due to various direct and indirect influences from humans. In order to prevent these crises we first need to understand the problem through academic study; we then need to put in place counter-measures to stem extinction and garner international cooperation of businesses and individuals. Meanwhile, in developing countries where nature is abundant, the reality is that relative to the utilization of their varied resources, Meanwhile, in developing countries where nature is abundant, the reality is that hardly any of the profit made in exploiting their abundant natural resources is spent on protecting its biodiversity. In order to avoid any further loss of diversity of wild animals, international agreements and coordination are indispensable.

References

1. Books

1) 小野泰洋，久保嶋江実：コウノトリ 再び，エクスナレッジ（2008），ISBN 978-4-767-80654-2
2) 山岸　哲 監，丁　長青 編著：トキの研究，エスプロジェクト（2007），ISBN 978-4-787-58566-0
3) 中村浩志：雷鳥が語りかけるもの，（ネイチャー・ストーリーズ），山と溪谷社（2006），ISBN978-4-635-23006-3
4) 長谷川博：アホウドリに夢中，（ノンフィクション 科学の扉）新日本出版社（2006），ISBN978-4-406-03244-5
5) 山岸　哲 編著：日本の希少鳥類を守る，京都大学学術出版会（2009），ISBN 978-4-876-98777-1
6) 山岸　哲 監，山階鳥類研究所 編：保全鳥類学，京都大学学術出版会（2007），

ISBN978-4-876-98703-0

2. Journal（2013 年 9 月現在）

1）http://www.ace-eco.org/issues/

3. URL（2013 年 9 月現在）

1）http://www4.ocn.ne.jp/˜ibis/（佐渡トキ保護センター）
2）http://www.stork.u-hyogo.ac.jp/（兵庫県立コウノトリの郷公園）
3）http://www.yamashina.or.jp/hp/yomimono/albatross/ahou_mokuji.html（アホウドリの復活への展望）
4）http://www.h7.dion.ne.jp/˜pb-eagle/（ワシ類鉛中毒ネットワーク）
5）http://www.env.go.jp/nature/yasei/hozonho/yanbarukuina.pdf（ヤンバルクイナ保護増殖事業計画）
6）http://www.pref.nagano.lg.jp/kankyo/hogo/kisyou2/20raicyo.pdf（ライチョウ保護回復事業）
7）http://www.umesc.usgs.gov/avian_conservation_ecology_team.html

Exercises

1．つぎの 1 ～ 71 の語に対応する英単語または英熟語を本文から選び出し，発音しなさい（動詞は原形を記入しなさい）。

	Japanese	English		Japanese	English
1	箱船		8	水田	
2	絶滅寸前		9	普及した	
3	絶滅		10	激烈な	
4	実施する		11	減退する	
5	皮肉にも		12	束縛	
6	ぐらつく		13	熱狂	
7	瀬戸際		14	なくなる	

	Japanese	English		Japanese	English
15	再導入する		36	森林減少	
16	剝製の		37	解剖	
17	標本		38	死骸	
18	保存する		39	水銀の	
19	死体		40	生殖力のない	
20	保護主義者		41	流産性の	
21	糧秣		42	未熟の	
22	捕食者		43	幼生	
23	群居性		44	トンボ	
24	巣立ちひな		45	孵化	
25	順応する		46	除去する	
26	悲劇		47	妊娠	
27	テン		48	おびき寄せる，おとり	
28	侵入する		49	住む	
29	果てしない		50	赤道	
30	肉食性の		51	翼長	
31	ツル		52	腹部の	
32	巣		53	禁止，追放	
33	雑食性の		54	汽船	
34	被食者，犠牲		55	植生	
35	湿地		56	及ぼす	

	Japanese	English		Japanese	English
57	踏みにじる		65	衛生	
58	原産の，土着の		66	消化不良	
59	地滑り		67	野生の	
60	移動させる（転座させる）		68	聖地，避難所	
61	啼鳴		69	摘む，淘汰する	
62	火山性の		70	危機	
63	地震		71	称揚する	
64	噴火				

2. つぎの各文が本文の内容と一致するものにはT(True)，一致しないものにはF(False) を，文末の（　　）に記入しなさい。

（1）Twenty-five percent of all endangered species are birds.（　　）

（2）The genetic backgrounds of the Japanese specimens of Japanese crested ibis, are not so different from the Chinese living individuals.（　　）

（3）Not like Japanese crested ibis, the oriental white stork have been maintaining its population.（　　）

（4）The Japanese brown frog（Rana japonica）lays eggs in Japanese summer season.（　　）

（5）Many creatures can maintain their populations by the delay of the mid-season drainage of rice paddy.（　　）

3. つぎの日本語の各文を（　　）の中の語を用いて英語の文にしなさい（必要があれば動詞を適切な形に変換しなさい）。

（1）生物多様性条約のもとで2020年までに種の絶滅の程度を半分にするように試みが行われているが，ほとんど効果がない。(attempt, implement, impact)

（2）21 世紀の終わりまでに，鳥類のおおよそ 30 % は絶滅すると推測されている。（die out, approximately, estimate）

（3）トキの個体の減少は，トキが農薬で汚染された餌を食べたためと考えられた。（decline, contamination）

（4）人工交配計画は 1956 年に開始された。（initiate, artificial）

（5）西洋では，コウノトリは赤ん坊を配達する幸運の生き物である。（creature, deliver, stork）

4. つぎの各問いに英語で答えなさい。

（1）What would be the reasons on the extinction of Japanese crested ibis in nature?

(2) How are the human-reared crested ibises returned for reintroduction to the nature?

(3) Do you know what a decoy is?

(4) Before Meiji era, why were those bird species maintained well?

(5) Which would be better to use the tax money for the protection of wild creatures or for supporting poor people?

Chapter 11
Battling by Deception：The Wondrous Mimicry of Creatures

When tending the garden in the summer, I occasionally notice a walking stick (insects of the order *Phasmatodea*). For a while, I watch in fascination at how it acts as if it were part of the branch. Watching it remain motionless even when the branch moves, and doing nothing but patiently staying still, I can't help but feel a sense of wonder for mimicry. In addition to walking sticks, a variety of creatures pretend to be what they are not, hiding from enemies or, conversely, lying in wait to attack. Indeed, it is a life-or-death battle.

In mimicry, the one that copies something else is called the mimic, and the emulated target is called the model. Depending on its purpose, mimicry can be divided into 1) camouflage, 2) standard mimicry, 3) aggressive mimicry, and 4) reproductive mimicry.

1) Camouflage: For example, as in the abovementioned walking stick or grasshopper, blending in with the surround by mimicking the patterns of branches or the ground to avoid being seen by predators.
2) Standard mimicry: A form of mimicry where the mimic makes itself more noticeable to indicate adverse consequences to predators. An example is the wing pattern of a non-poisonous butterfly that resembles that of a poisonous butterfly.
3) Aggressive mimicry: A means of assimilation for the purpose of capturing prey. The tactic is the same as that of camouflage. A butterfly is captured by a praying mantis holding onto a twig that it mimics.
4) Reproductive mimicry: Although relatively rare, mimicry for

the purpose of reproduction exists. For example, a species of orchid known as the Australian hammer orchid (*Drakaea spp.*) mimics a female wasp and makes the congregating male wasps serve as pollinators.

Depending on the content, mimicry can also be classified into 1) Batesian mimicry and 2) Mullerian mimicry.

1) Batesian mimicry: Mimicry in which a poisonous creature is emulated. For example, the flying style, buzzing, and body shape of the hoverfly (*Milesia undulata*) are very similar to those of a wasp. She momentarily deceives the opponent and escapes while it hesitates.

2) Mullerian mimicry: For example, poisonous butterflies have similar wing patterns even though the species may differ. This is presumably not only to signal to the predator that they are poisonous, but also to reduce the risk of predation by emulating one another. A characteristic feature of Mullerian mimicry is that both the mimicking and mimicked are poisonous.

Having distinguished the types of mimicry, let us next consider a number of characteristic examples of mimicry (**Figure 11**).

Butterfly mimicry : There are quite a few "bug mongers" who love insects and excel at collecting them. Among them, butterflies are by far the most popular insects. Some butterflies are either poisonous or taste bad to predators. There are also many non-poisonous butterflies that mimic such poisonous butterflies. The most characteristic mimicry is that of the wing pattern. Poisonous butterflies have a colorful pattern to signal to predators that they are poisonous. For example, one poisonous butterfly called the common

---

Notes : Australian hammer orchid (*Drakaea spp.*)「西オーストラリア原産のランの一種」
hoverfly (*Milesia undulata*)「シロスジナガハナアブ」
common rose butterfly (*Pachliopta aristolochiae*)「ベニモンアゲハ」

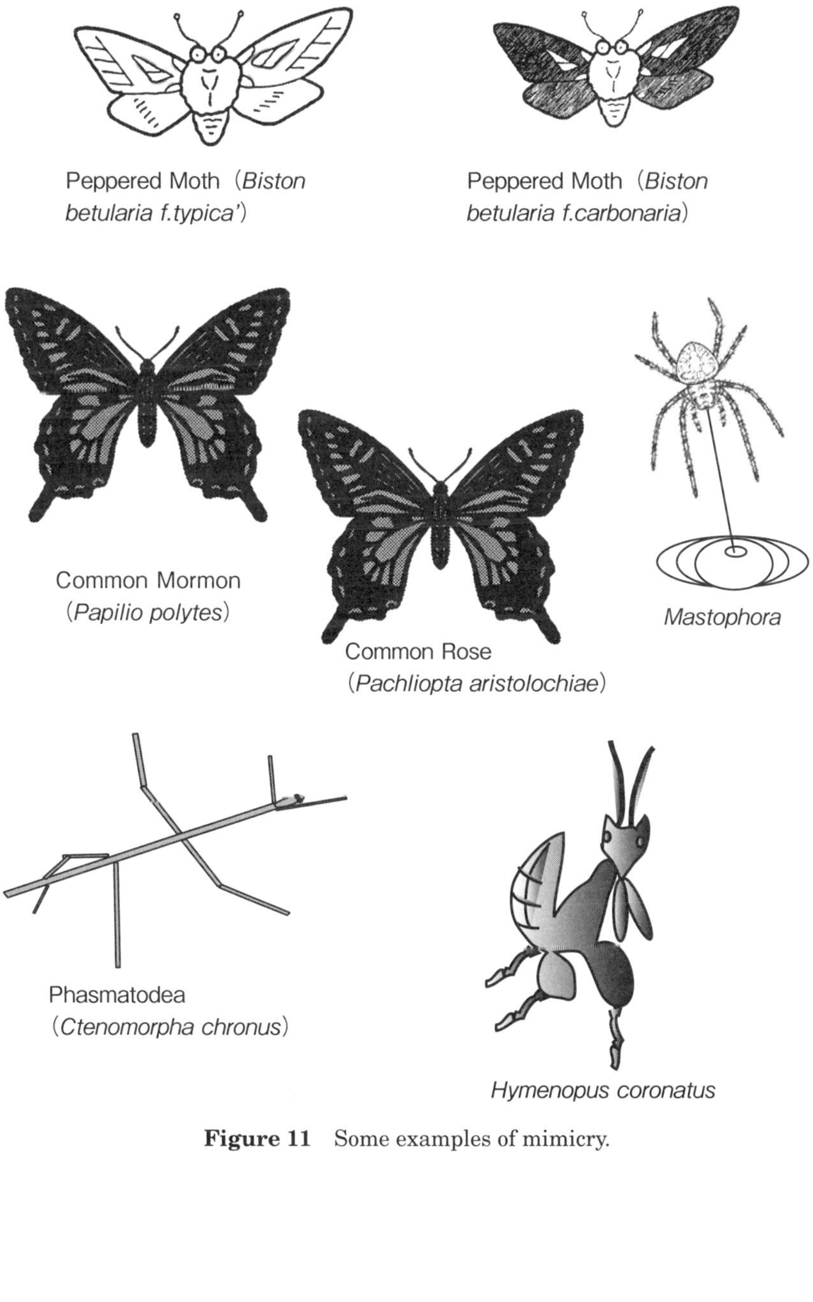

Figure 11 Some examples of mimicry.

rose butterfly (*Pachliopta aristolochiae*) has a pattern of white and orange spots on a black background. These warning colors, which stand out, visually inform the predator that the butterfly is poisonous. And there is a butterfly that is alike it in every way: the non-poisonous common mormon butterfly (*Papilio polytes*). So this butterfly is a mimic of the poisonous common rose, and is a typical example of a Batesian mimic.

Butterflies are known as mimics not only as adults but from the larval stage as well. In the early stages, the larva of a Spicebush Swallowtail (*Papilio troilus*) has a brown and white body surface, almost like a bird dropping. Predators of caterpillars, mainly birds, that see this larva appear to mistake it for a bird dropping. In this way, the larva can avoid predation. However, in its final stage, the larva no longer has the black-and-white dropping pattern; rather it changes to a camouflaging color resembling that of grass, eating leaves while hiding among the green at the same time. It is a very strange story. It is becoming clear that the changes in the larva are in fact controlled by juvenile hormone (JH). This Spicebush Swallowtail molts four times before becoming a chrysalis (complete metamorphosis or holometabolism). The JH concentration is high until the 4th instar larva. However, the JH concentration decreases thereafter, and the body color changes to a green color, completely different from the previous color, in the final 5th instar larva. The changes are not limited to these effects, as the structure of surface protrusions and pigment distributions of the eyespot are also adjusted. So, what happens if 4th instar larva is treated with JH? As the JH concentration remains high, the final 5th instar larva becomes black

---

Notes : common mormon butterfly (*Papilio polytes*)「シロオビアゲハ」
Spicebush Swallowtail (*Papilio troilus*)「クスノキアゲハ」

and white. If the JH treatment of the 4th instar larva is delayed, then a 5th instar larva with an intermediate pattern (the bird-dropping pattern and cryptic green coloration) is born. These results revealed that the patterns of gene expression are switched in unison depending on the JH concentration.

Melanism : evolutionary change by human activities

Melanism in moths is also an example of mimicry. In fact, it is also an example of how melanism is intimately related to human activity. Many of you may remember having learnt of this phenomenon, which has been called "industrial melanism", in high school textbooks. In England, where the industrial revolution began in 1800, factory towns became polluted with soot because coal was used in massive quantities as a power source. As a result, the lichen that had grown on the surface of trees died out from air pollution and the trees changed to a blackish color. The peppered moth (*Biston betularia*) is a common moth seen everywhere in England. The wings of this moth turned black for individuals living in industrial regions. This melanism constitutes cryptic coloration in which the moths are camouflaged by the blackened trees. In contrast, peppered moths that were originally white began to stand out in the city because of their whiteness, and were found easily by birds and eaten. Thus in metropolitan areas, white moths declined and black moths came to predominate. In fact, it was only in 1950 that this fact was experimentally verified. White and black moths were released in an industrial area or rural area. After a few days, the moths were collected with a light trap and their numbers were examined. The

---

Notes : industrial melanism「工業暗化」
peppered moth (*Biston betularia*)「オオシモフリエダシャク」

survival rate for black moths in the industrial area was higher than that of white moths, whereas in the rural areas the survival rate of white moths was higher than that of black moths. The peppered moth was originally white. The dark type is caused by the presence of a single dominant mutation. This industrial melanism is assumed to be a change (or evolution) caused by selection in the context of the phenomena of mutants and air pollution.

Up to this point, we have examined mimicry for avoiding detection by enemies (camouflage). Let us next examine mimicry for capturing prey (aggressive mimicry).

The orchid mantis (*Hymenopus coronatus*) is a praying mantis that lives in India and Southeast Asia. When it is perched on an orchid, distinguishing the mantis from the flower is quite difficult unless one looks very carefully. The mantis is patterned with thin green lines on a white and pink background. Through mimicry, it protects itself from natural predators and simultaneously captures prey that approaches it as a flower. If you look only at the orchid mantis, you would probably realize it to be a praying mantis. However, when on a branch amid orchid flowers, it is so beautiful that you might call it a work of art. It is not only the color that is deceptive; the animal also assumes unique poses for deception. However, its movements are slow, and it would be eaten quickly if a predator discovered it. Furthermore, as moving would give it away as a mimic, all it can do is to wait patiently for a prey to approach. As a small larva, it also mimics a stinkbug — an insect that releases an odor. This is a classical type of Batesian mimicry.

In this way, visual mimicry is by far the predominant form. How then is the visual appearance of a mimic adjusted to approximate that

---

Note : orchid mantis (*Hymenopus coronatus*)「ハナカマキリ」

of the model? Let us take as an example a butterfly that uses visual mimicry. These studies have primarily been carried out by American scientists. Currently, we understand the formation of eyespots on butterfly wings. During the larval period, butterflies and moths internally form a small area of undifferentiated tissue known as the wing disc. During the initial or prepupal stage of metamorphosis, this is stimulated by insect hormones called ecdysteroids and undergoes cell division and tissue differentiation. Furthermore, in the pupal stage, apoptosis and scale formation occur in the area around the wing disc. Eyespot formation is actually controlled by *Dll* (distal-less), the same gene that controls leg formation. Although wings and legs may appear different at first glance, their structures are very similar. Once the outline of the eyespot is determined by the expression of *Dll*, the task of coloring the pattern is carried out next. Pigments and proteins secreted by the cells solidify to form small colored fragments called scales, which in turn produce the patterns. Unfortunately, only eyespots are understood thus far among butterfly patterns.

Mimicry based on non-visual information exists as well. The bolas spider (*Mastophora*) hangs from scaffolding-like threads that it spins from the tips of its legs. With its front legs, the spider then spins a thread with a globule of viscous fluid at the tip. This "bolas" contains a substance resembling the sex pheromone of moths. Lured by this pheromone, the male moth arrives and is captured and eaten by the bolas spider. This is an example in which a chemical substance is used for mimicry. In addition, the titan arum (*Amorphophallus titanum*), which releases a putrid stink, is a plant that uses olfaction to attract pollinating flies that uses smell to attract flies. This too is a form of

---

Notes : bolas spider (*Mastophora*)「ナゲナワグモ」

titan arum (*Amorphophallus titanum*)「スマトラオオコンニャク」

mimicry. Members of the firefly family exchange information with light signals. In the genus *Photuris*, the male emits a pattern of light signals that mimics the signal of other species and preys on the females that congregate. This constitutes predation based on mimicry by light.

In the case of Batesian mimicry, the association between the mimics and the model requires that the predator, after attacking the model obstacle to the predator, the predator retains a memory of the hazard and then learns. Otherwise, the predator would attack the mimic prey without any hesitation. Retaining a memory is only effective for predators in which neurons and vision are developed to some degree. The number of mimics is also limited. If it is greater than the number of models (harmful), although the predator may recognize that a mimic may be poisonous from warning colors depending on its experience, the likelihood that it is harmless will increase. In this case, the warning colors will lose all meaning. Thus considered, the predators need to reproduce so that their numbers do not exceed the number of models. Without doubt, balancing the number of mimics is difficult.

References

1. Books

1) 藤原晴彦：似せてだます擬態の不思議な世界（DOJIN 選書 2）化学同人（2007），ISBN978-4-759-81302-9

2. Journal

1) The Heliconlus Genome Consortium 2012 : Butterfly Genome Reveals Promiscuous Exchange of Mimicry Adaptations among Species. Nature doi:10.1038/nature11041

Note : *Photuris*「ホタルの仲間」

3. URL（2013 年 9 月現在）

1）http://www.idensystem.k.u-tokyo.ac.jp/research.htm

Exercises

1. つぎの 1 ～ 63 の語に対応する英単語または英熟語を本文から選び出し，発音しなさい（動詞は原形を記入しなさい）。

	Japanese	English		Japanese	English
1	欺くこと		17	だます	
2	すばらしい		18	躊躇（ちゅうちょ）する	
3	擬態		19	虫	
4	ナナフシ		20	商人	
5	枝		21	勝る	
6	まねる		22	幼生の	
7	偽装		23	脱皮	
8	バッタ		24	蛹	
9	囲む		25	変態	
10	捕食者		26	完全変態	
11	被食者		27	齢	
12	同化		28	色素	
13	策略		29	隠れた / 秘密の	
14	カマキリ		30	現す	
15	小枝		31	遺伝子発現	
16	集める		32	調和	

	Japanese	English		Japanese	English
33	眼状斑点		49	断片	
34	黒化		50	分泌する	
35	汚染する		51	凝固させる	
36	地衣類		52	紡ぐ	
37	卓越する		53	球体	
38	実証する		54	粘った	
39	現象		55	足場	
40	～の真ん中に		56	誘惑	
41	欺くような		57	フェロモン	
42	カメムシ		58	腐敗した	
43	未分化の		59	悪臭を放つ	
44	前蛹		60	嗅覚	
45	エクジステロイド		61	放射する	
46	蛹の		62	障害	
47	アポトーシス		63	危険	
48	鱗片				

2. つぎの各文が本文の内容と一致するものには T(True)，一致しないものには F(False) を，文末の（　　）に記入しなさい。

（1）An insect species, walking stick has a mimicry capacity.（　　）

（2）In the rural areas the survival rate of white moths was higher than that of black moths.（　　）

（3）Visual mimicry is by far the predominant form of the mimicry.（　　）

（4）Orchid mantis resembles an orchid flower.（　　）

（5）Camouflage is not the common type of mimicry.（　　）

3. つぎの日本語の各文を（　　）の中の語を用いて英語の文にしなさい（必要があれば動詞を適切な形に変換しなさい）。

（1）工業地域では，黒い蛾の生存率は，白い蛾より高かった。(industrial, survival rate)

（2）偽装によって敵の捜索から逃れる擬態について調査した。(examine, camouflage, detection)

（3）数日後に蛾は誘蛾灯で収集され，個体数が数えられた。(light trap, examine)

（4）蝶は成虫だけではなく幼虫期にも擬態することが知られている。(mimic, adult, larval stage)

（5）*Drakea* 属のランは雌のスズメバチに擬態し，集まってくるオスのハチを授粉者として使う。(congregating, pollinator)

4. つぎの各問いに英語で答えなさい。

（1）What types of mimicry, can you classify?

(2) What is an advantage of mimicry in nature?

(3) Is mimicry an acquired phenotype by learning or under genetic control?

(4) What has happened in England at the 19th century?

(5) How can the bolas spider mimic?

Discuss Corner 3

Technologies Inspired by Nature

The natural world is intrinsically endowed with many sophisticated structures and functions that are difficult for humans to emulate. Biomimetics, otherwise known as biomimicry, is a field of a new academic discipline, which aim is to analyze and emulate these structures and functions and incorporate them into technologies in human society. As we will see, some surprising creatures are endowed with a variety of structures and functions.

参 考

http://ngm.nationalgeographic.com/2008/04/biomimetics/tom-mueller-text
http://news.discovery.com/adventure/tags/biomimetics.htm

Discuss how we can learn from the nature for technology applications for betterment of the human life.

Chapter 12
What the Medaka (*Oryzias latipes*) Genome Tells Us?

Model organisms form the foundations of biology. Well-known models include the mouse and fruit fly. Additional models include the chimpanzee and rhesus macaque (primates), African clawed frog (amphibian), zebrafish (fish), rice and *Arabidopsis thaliana* (plants), sea squirts, koji (*Aspergillus oryzae*, a mold), and even bacteria. Amid this backdrop, the genome of the medaka is currently attracting attention.

The medaka is a special fish for the Japanese. Until very recently, if you looked carefully, you could find them in rice paddies and brooks. The medaka is about 4 cm long with eyes that are large for its body size giving it a humorous countenance. It lives throughout Japan except Hokkaido, as well as on the Korean peninsula, China, and even Taiwan.

Medaka live in freshwater or brackish water and cycle through a number of generations from spring to summer. In the winter they hibernate. Their lifespan is approximately 1 to 3 years. They are easy to raise and hardly require any attention to water temperature or oxygen concentration. They are omnivorous and can be raised on either zooplankton or phytoplankton. Sexing is simple, as animals with the larger dorsal and tail fins are the males and those with the larger pectoral and pelvic fins are the females. During the spawning period, the female is sometimes seen with eggs adhered to her abdomen.

Notes：*Arabidopsis thaliana*「シロイヌナズナ」, *Aspergillus oryzae*「麹菌」

The medaka has long been used as an experimental animal in Japan. However, it began to attract worldwide attention after its genome was sequenced in 2007. Until then, only the zebrafish, another model organism, had been at the center of genomic research.

Classification of sex in the medaka

Sexual differentiation of the medaka is what first attracted attention. The sex of a medaka can easily be distinguished even by a layperson. Similar to humans, the medaka has one set of sex chromosomes. Their combinations are also the same as that of humans, with XY becoming male and XX becoming female (in other words, the sex is determined by the sex chromosome X). How is the sex of a medaka specifically determined?

Normally, a fertilized ovum divides in the process of sexual differentiation, proliferates, and becomes a germ cell (**Figure 12**). This germ cell moves to the center of the embryo and forms the genital primordium. Subsequently, it differentiates into a female-type cell and forms the ovaries. These ovaries produce estrogen, a sex steroid hormone, which acts on the entire body and forms the female-type pectoral and pelvic fins.

Meanwhile, the gonadal somatic cells created after germ cell formation also form a genital primordium, which similarly undergoes sexual differentiation to produce the testes. These testes produce testosterone, a sex steroid hormone, which acts on the entire body to form the dorsal and tail fins unique to the male. If a microscope that could visualize the living embryo as is existed, sexing would already become possible from the 8th day after fertilization. With further growth, sexing becomes possible for anyone.

Thus in the medaka, after the gonadal sex is determined, the difference in body form between the sexes is determined by the action

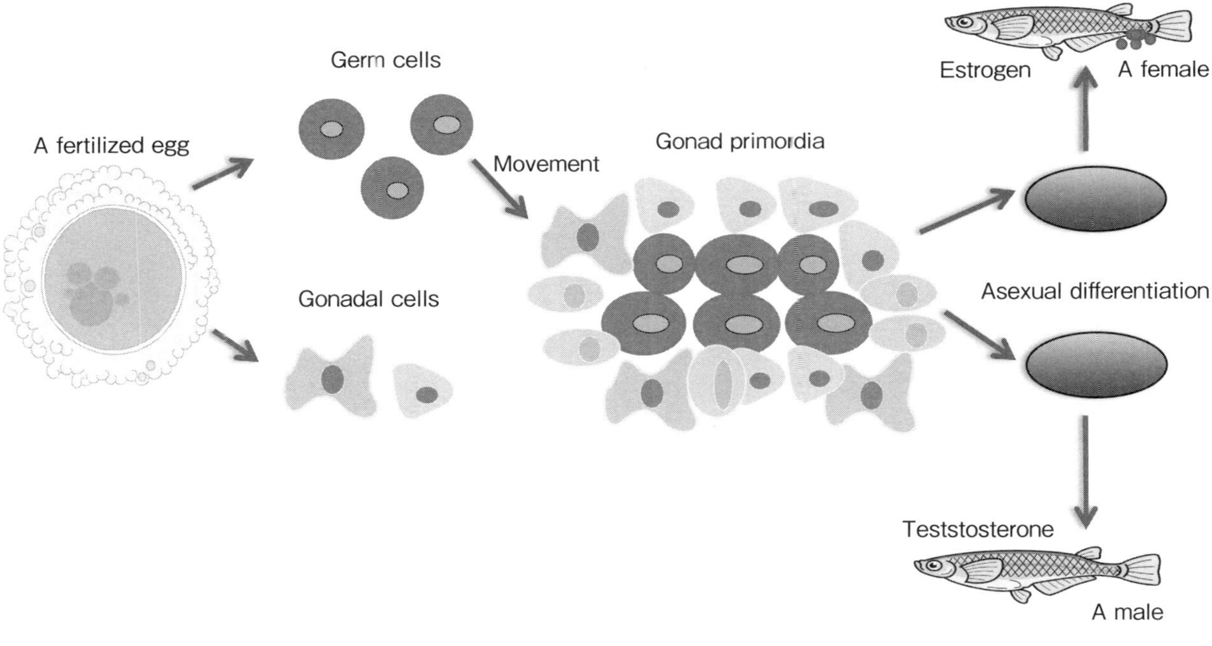

Figure 12 The sexual differentiation of Japanese medaka (*Oryzias latipes*), modified from Nature DIGEST.d

of hormones. The actions of the gonads and germ cells are the major keys to this process. It was found that each was formed at a different place in the embryo, and that the germ cells migrated and became a complete gonad by mutual interactions between the cells.

Next, whether the medaka would become male or female was tested by blocking the migration of germ cells. The resultant medaka could produce neither ova nor sperm, although its body characteristics were completely masculinized. In other words, regardless of the sex chromosome combination, the somatic cells of a medaka lacking germ cells are all masculinized. That is, gene expression, hormone-producing endocrine cells, and the hormones throughout the body are all masculinized. This suggests that germ cells play an important role in gonadal masculinization. Medaka lacking germ cells were masculinized, and the estrogen-producing endocrine cells characteristic of the female were found to be determined not by the sex chromosomes but by the germ cells. These findings overturned results obtained in mice.

Evolution of color vision

Important findings on the evolution of color vision are on the verge of being revealed by research on the medaka. Humans perceive objects as trichromatic (red, green, and blue) color images. However, mammals generally have dichromatic vision. It is thought that primates (as represented by humans) came to live in treetops to avoid predators, and thus developed trichromatic vision, which allows good vision in sunlight as well. What about other creatures?

Although it may seem surprising, birds and insects have tetrachromatic (red, green, blue, and ultraviolet (UV)) vision. Fish see even more colors.

Specifically, the role of the protein known as "opsin", which distinguishes color, is being investigated. An examination of the role of opsin in zebrafish revealed a total of eight opsin genes: two red-type, four green-type, one blue-type, and one UV-type (humans have only three types). This implies that fish distinguish what we see with many more colors.

Seeing objects in water is quite difficult compared to seeing objects on land because the light environment of water changes greatly owing to the effects of depth, turbidity, and water surface fluctuations. Fish have adapted in this way to a variegated aqueous environment, and their color vision has evolved in complex ways.

The differentiation, retrogression, and development of vision that can distinguish color have important implications for the evolution of organisms. How do these genes for color vision vary with different environments? Studying this in zebrafish is difficult because unlike medaka, hardly any lineages originating from different environments have been discovered. Therefore, based on the information on opsins obtained from zebrafish, the medaka genes were examined. This revealed, for instance, that one green-type opsin was present. This type has also been found in a cichlid (fish) but not in zebrafish. This gene can be viewed to have undergone genetic duplication at a very early point in evolution, and to have thus moved into various lineages of fish. The findings could provide important clues to the evolution of color vision.

What sort of research will the medaka, whose entire genome sequenced in 2007, contribute to in the future?

The vertebrates include fish, amphibians, reptiles, birds, and mammals. Although their appearances and forms differ at first glance,

---

Notes：cichlid「シクリッド，カワスズメ」淡水に生息する魚類

a general cross-species mechanism by which a rough form is produced from a single fertilized egg is said to exist. In other words, the deciphering of the medaka genome is expected to lead to clues for understanding universal developmental principles for vertebrates including humans.

Among model organisms, the entire genomes of humans and a variety of other organisms are being deciphered. A comparison of the human and medaka genomes shows that 60 % of the genes are shared, and that this proportion even increases to 80 % when similar genes are included. Furthermore, a disease shared between medaka and humans has been discovered in recent years. By studying and analyzing this disease, we should be able to understand its origins and mechanisms. In addition, this should also play an important role in the field of drug development.

The aforementioned discovery of sexual differentiation in the medaka also helps understand sexual differentiation in different organisms. For example, the Y chromosome of the human and medaka determines sex. However, in the nematode and fruit fly, for example, it is determined by the quantitative ratio of sex chromosomes to normal chromosomes. In the silkworm, the presence or absence of the W chromosome determines sex. Although there are various cascades leading to male- or female-type animals, many of the details remain unclear. Results from the medaka genome will also be highly informative for such research.

In the environmental field, the medaka is contributing to studies that are exploring the effects, both longitudinally and across many generations, of substances with the potential to cause bodily changes. The amazing medaka is likely to continue to rise in status as a model organism in the future.

References

1. Books

1）吉川　寛，堀　寛 編：研究をささえるモデル生物―実験室いきものガイド―，化学同人（2009），ISBN978-4-759-81178-0

2. Journal

1）M. Kasahara, et al.：The Medaka Draft Genome and Insights into Vertebrate Genome Evolution. Nature 447：pp.714-719（2007）
2）JH. Postlethwait, et al.：Vertebrate Genome Evolution and the Zebrafish Gene Map. Nature Genetics 18：pp.345-349（1998）
3）G. Rubin, et al.：Comparative Genomics of the Eukaryotes. Science 287：pp.2204-2215（2000）

3. URL（2013 年 9 月現在）

1）http://www.shigen.nig.ac.jp/medaka/genome/top.jsp

Exercises

1．つぎの 1 ～ 66 の語に対応する英単語または英熟語を本文から選び出し，発音しなさい（動詞は原形を記入しなさい）。

	Japanese	English		Japanese	English
1	ゲノム		7	アフリカツメガエル	
2	モデル生物		8	ゼブラフィッシュ	
3	ショウジョウバエ		9	ホヤ	
4	アカゲザル		10	背景	
5	霊長類		11	小川	
6	両生類		12	こっけいな	

	Japanese	English		Japanese	English
13	表情		34	生殖原基	
14	汽水		35	胚	
15	冬眠する		36	卵巣（子房）	
16	寿命		37	エストロゲン	
17	雑食性の		38	ステロイド	
18	動物プランクトン		39	生殖腺体細胞	
19	植物プランクトン		40	テストステロン	
20	尾ひれ		41	顕微鏡	
21	背びれ		42	卵子（複数）	
22	胸びれ		43	精子	
23	腹びれ		44	雄性化させる	
24	産卵 / 卵		45	内分泌	
25	くっつく		46	ひっくり返す	
26	腹部		47	知覚する	
27	素人／俗人		48	三色の	
28	染色体		49	哺乳動物	
29	受精する		50	二色の	
30	卵子		51	こずえ	
31	性分化		52	生き物	
32	生殖細胞		53	濁り，攪乱	
33	精巣		54	変動	

	Japanese	English		Japanese	English
55	まだらの		61	手がかり	
56	水の		62	脊椎動物	
57	退化		63	解読する	
58	含蓄		64	カイコ	
59	血統		65	つながり	
60	遺伝的重複		66	縦の	

2. つぎの各文が本文の内容と一致するものにはT(True)，一致しないものにはF(False）を，文末の（　　）に記入しなさい。

（1）Medaka is omnivorous.（　　）

（2）Medaka dose not live in Hokkaido.（　　）

（3）Medaka does not hibernate in Japan.（　　）

（4）Medaka requires specific cultural conditions.（　　）

（5）Medaka is a new model organism.（　　）

3. つぎの日本語の各文を（　　）の中の語を用いて英語の文にしなさい（必要があれば動詞を適切な形に変換しなさい）。

（1）メダカの寿命は概ね１～３年である。(lifespan, approximately)

（2）メダカゲノムが解読されるまで，ゼブラフィッシュがゲノム研究の中核的モデル生物であった。(decipher, model organism, genome)

（３）メダカの性分化はまず最初となる興味深い課題である。(sexual differentiation, attention)

（４）ヒトとメダカのゲノムを比較すると，遺伝子の 60 % は類似している。(comparison, resemble, show)

（５）脊椎動物には，魚類，両生類，は虫類，鳥類およびほ乳類がある。(include, vertebrates)

4. つぎの各問いに英語で答えなさい。

（１）How is the sex of a medaka specifically determined?

（２）Why is medaka regarded as an important model organism?

（３）Can fish recognize many different colors?

（４）How would the vision associate with the evolution?

(5) What sort of research will the medaka, whose entire genome sequenced in 2007, contribute to in the future?

Discuss Corner 4

Potential Synthesis of Living Organism

Artificially synthesized genomes are reality in the present world. Research based on this type of method has been advanced with mycoplasma, a genus of prokaryotic bacteria. However, the length of its genome — 582, 970 bases — is short. Given that this synthetic mycoplasma comprises a group of genes and divides, it could be considered a life form.

参 考　DG. Gibson：Complete chemical synthesis, assembly, and cloning of a *Mycoplasma gentitalium* Genome. Science **309**, pp.1215-1220 (2008)

Discuss how you see the artificial creation of the life form.

Discuss Corner 5

The Mammoth Resurrection Project

In the opening scene of *Jurassic Park*, the motion picture based on the 1990 novel of the same title, dinosaur genes are collected from the blood of a mosquito that had been trapped in amber. These were analyzed and repaired, any missing sections were supplemented with frog DNA, and the genes were then inserted into unfertilized crocodile ova. The story is about dinosaurs that were supposed to have died out come back to life. Actually, there are many places in the movie that scientifically do not add up. Nonetheless, people were excited by the spirit of romantic adventure in "dinosaur resurrection", and the movie became a blockbuster.

Around the time of the original work, scientists aiming to resurrect extinct creatures were beginning to emerge. Their target was not a dinosaur but a mammoth. Extinct mammoths were being discovered one after another in the Siberian tundra, and scientists scrambled to use the genetic information from these mammoths to somehow resurrect a mammoth.

参 考　W. Miller, et al.：Sequencing The Nuclear Genome of The Extinct Woolly Mammoth. Nature **456**：pp.387–392 (2008)
http://phenomena.nationalgeographic.com/2013/03/12/the-promise-and-pitfalls-of-resurrection-ecology/

Discuss the possibilities of science and technology for the aim of the mammoth resurrection such as genetic compatibility and reproductive biology questions, and also debate on the ethics of the resurrection.

Answers to Exercises

Chapter 1

Exercises 1

1. planarian, 2. produce, 3. organism, 4. lizard, 5. newt, 6. human, 7. scratch, 8. bleed, 9. stem cell, 10. aggregate, 11. replenish, 12. disease, 13. injury, 14. restoration, 15. artificial, 16. create, 17. fibloblast, 18. mouse (mice), 19. differentiate, 20. cardiac muscle, 21. cluster, 22. embryo (nic), 23. fertilize, 24. immunological rejection, 25. catalyst, 26. nucleus (nuclei), 27. mature, 28. identical, 29. uterus, 30. genetic, 31. clone, 32. overturn, 33. organ, 34. rewind, 35. manipulation, 36. transcription, 37. transplant, 38. neural, 39. insertion, 40. culture, 41. regeneration, 42. treatment, 43. udder cell, 44. spinal, 45. symptom, 46. diabete, 47. proliferate, 48. pathogenesis, 49. drug, 50. administer, 51. Rejection response, 52. toxicity

Exercises 2

(1) T, (2) F, (3) F, (4) T, (5) F

Exercises 3

(1) A Japanese scientist generated the world's first iPS cells (induced pluripotent stem cells) from mice.

(2) Scientists engaging in medical sciences must always make ethical consideration on the personal information of patients.

(3) The birth of Dolly the sheep in 1996, made the systematic generation of iPS cells.

(4) Transplanting neural stem cells derived from iPS cells into mice with artificially given spinal injuries reduced the symptoms by 20 %.

(5) By transforming the iPS cells into the targeted tissue, the effects and toxicity of a new drug candidate can be examined.

Exercises 4

(1) It may be debatable in ethical view points, however, it is a great opening in supporting medical research.

(2) Go to the text statement.
(3) Four planarians regenerate.
(4) Yes / No up to your thought.
(5) Debates can be made on ethical legal and social implications.

Chapter 2

Exercises 1

1. extermination, 2. bit, 3. parasite, 4. infect, 5. nausea, 6. salivary gland, 7. vector, 8. bloodstream, 9. penetrate, 10. hepatocyte, 11. multiply, 12. asexual, 13. reproduction, 14. sexual, 15. gametocyte, 16. ingest, 17. circulate, 18. infant, 19. overwinter, 20. countermeasure, 21. span, 22. clinical trial, 23. unprecedented, 24. pertain, 25. thread, 26. insecticide, 27. ingredient, 28. insect-repelling, 29. innocuous, 30. greenhouse, 31. alight, 32. popularize, 33. subsidy, 34. efficacy, 35. render, 36. manipulate, 37. debate, 38. eradicate, 39. dawn, 40. verge

Exercises 2

(1) F, (2) F, (3) T, (4) F, (5) T

Exercises 3

(1) If a person becomes infected by malaria, symptoms such as fever, headache, and nausea occur.
(2) There are other methods to prevent malaria infection as well.
(3) With the spread of malaria vaccines and mosquito nets, or with methods based on genetic manipulation, the day that malaria is eradicated may not be far off.
(4) This medicine was administered to many adult men, and the clinical trial is on an unprecedented scale.
(5) Prevention and control measures supported by WHO, have led to a reduction in malaria mortality rates by more than 25 % globally since 2000.

Exercises 4

(1) The malaria parasite is present in the salivary glands of *Anopheles* mosquitoes.
(2) It is an artificial sterility characteristics.
(3) It reduces bites by mosquitos, consequently the risk of the infection can be decreased.

(4) *Chrysanthemum cinerariaefolium* can provide insect repellants. Quinine from *Cinchona calisaya* trees, is the most effective medicine yet and *Artemisia* species can be used for curing as a herbal medicine.
(5) No, many different efforts by various scientific disciplines, should be amalgamated such as parasitology, entomology, ecology, agro-chemistry, pharmacology, medical science, rural sociology, etc.

Chapter 3

Exercises 1

1. influenza, 2. pandemic, 3. ancient, 4. smallpox, 5. tubercurosis, 6. syphilis, 7. cholera, 8. widespread, 9. antibiotic, 10. incurable, 11. virus, 12. decline, 13. efficacy, 14. drug, 15. novel, 16. avian, 17. casualty, 18. proliferate, 19. raise, 20. virulence, 21. estimate, 22. mortality, 23. underlie, 24. pregnant, 25. elderly, 26. classify, 27. internal, 28. protein, 29. trigger, 30. symptom, 31. epidemic, 32. emergence, 33. discover, 34. immunity, 35. fact, 36. survive, 37. structure, 38. subtype, 39. insert, 40. membrane, 41. envelope, 42. circulate, 43. designate, 44. combination, 45. shift, 46. drift, 47. re-aggregation, 48. gene, 49. mutation, 50. identity, 51. breakneck, 52. fusion, 53. swine, 54. spread, 55. wildfire, 56. property, 57. transportation, 58. expansion, 59. respiratory, 60. gastrointestinal, 61. cough, 62. diarrhea, 63. pneumonia, 64. myocarditis, 65. dysfunction, 66. brain, 67. encephalopathy, 68. description, 69. fate, 70. eternal

Exercises 2

(1) F, (2) T, (3) F, (4) T, (5) F

Exercises 3

(1) The battle against infectious diseases will come to the end in a near future.
(2) The swine influenza had cases on infection also to people and caused casualties.
(3) A virus with low virulence might mutate to a more virulent one is also possible.
(4) RNA virus can undergo its genetic change more easily than a DNA virus.
(5) The war between people and viruses has been fated to continue

eternally.

Exercises 4

(1) Yes, it is a very challenging job but one may need a lot of precaution to handle with human infectious viral agent.

(2) Influenzas may be one we can think about easily as we hear so many transmission incidences of avian and swine influenzas to mankind. Also with more access of humans to wild-lives in nature, we should be careful on more possibilities of zoonosis.

(3) There may be four or more key components : human exposure to the virus vector such as a bird, bird exposure to the virus, rate of the change to more virulence form and rapid mutation of the virus to infect humans.

(4) Viral diseases are more challenging entities as the changes in severe virulence and host ranges are so fast and variable.

(5) Knowledge on prevention and hygiene is cardinal and yet we need medical research for the direct combats with the viruses.

Chapter 4

Exercises 1

1. Sense, 2. neuroscience, 3. acquisition, 4. vision, 5. function, 6. impaired, 7. pronounce, 8. constitutive, 9. blindness, 10. cataract, 11. prevalence, 12. glaucoma, 13. muscular, 14. degeneration, 15. trachoma, 16. parasitic, 17. lens, 18. implant, 19. decline, 20. transplantation, 21. donate, 22. neuron, 23. neuroplasticity, 24. retina, 25. defect, 26. optic, 27. invisible, 28. deform, 29. rehabilitation, 30. perceive, 31. endow, 32. distinguish, 33. rotate, 34. eyebrow, 35. skill, 36. dimension, 37. engage, 38. property, 39. discard, 40. plastic, 41. detrimental, 42. restore, 43. connection, 44. recover

Exercises 2

(1) F, (2) T, (3) F, (4) T, (5) F

Exercises 3

(1) The causes of visual impairment can be divided broadly into two types.

(2) In recent years, lens implants have become facile to perform in developed countries owing to technological progress.

(3) If there is any defect in the optic nerve, objects may appear to be

invisible, partially invisible, or deformed even if the eye itself is perfectly normal.

(4) Individual neurons are connected to other neurons to form a network of neurons.

(5) There are reports that vision recovers considerably by continuing rehabilitation after corneal transplantation.

Exercises 4

(1) It may be true but can be a scientific fiction.

(2) The causes of visual impairment can be divided broadly into two types : acquired and genetic or constitutive causes.

(3) It is my great interest. Or I am not on such a research subject?? Show your interest?

(4) Participatory agreements shall be set among the stakeholders such as patients, physicians, regulators of the medical technology including ethical view points.

(5) 1) The ability to perceive faces
 2) The ability to perceive depth
 3) The ability to recognize objects

Chapter 5

Exercises 1

1. disorder, 2. phenomenon (phenomema), 3. reclaim, 4. comfortable, 5. night-oriented, 6. awake, 7. stabilize, 8. memorize, 9. perform, 10. erroneous, 11. retrieval, 12. ingest, 13. act, 14. witness, 15. testify, 16. substantial, 17. napping, 18. release, 19. imply, 20. feature, 21. terrify, 22. strangle, 23. wakefulness, 24. paralysis, 25. disruption, 26. scattered, 27. bout, 28. underlie, 29. examination, 30. facilitate, 31. syndrome, 32. concentration, 33. depressive, 34. obstructive, 35. abdominal, 36. thoracic, 37. oronasal, 38. impotence, 39. respiratory, 40. diagnose, 41. apnea, 42. hypopnea, 43. affliction, 44. sagging, 45. jawbone, 46. episode, 47. protrude, 48. skeletal, 49. tonsil, 50. throat, 51. ailment, 52. hypertension, 53. arrhythmia, 54. mouthpiece, 55. indicator, 56. anxiety, 57. outpatient clinic, 58. checkup

Exercises 2

(1) F, (2) T, (3) F, (4) T, (5) F

Exercises 3

(1) Before the experiment, the subjects were asked to memorize four simple terms that were related.
(2) This false memory occurs frequently in groups that do not get adequate sleep during the night.
(3) A characteristic of the OSAS, is a thickening of the neck from obesity.
(4) Obstructive sleep apnea (OSA) is diagnosed when the apnea/hypopnea index is at least 5 or more.
(5) For nearly every case of narcolepsy, the onset occurs in the teens to 30s, and the symptoms often become milder with age.

Exercises 4

(1) The first step starts with the patient fully understanding that he or she is a narcoleptic. The second is to avoid heavy drinking or irregular lifestyles and not to accumulate stress.
(2) There are several features : Breathing stops for 10 seconds or more (apnea), lack of concentration, heavy-headedness or discomfort on awakening, depressive feelings, and impotence.
(3) SAS could have interactions such with hypertension, arrhythmia, ischemic heart disease, and diabetes.
(4) The treatment considered most effective is weight reduction.
(5) One could say that this shows the extent to which sleep is an important indicator of health.

Chapter 6

Exercises 1

1. overwhelming, 2. allergic, 3. cedar, 4. pollen, 5. drift, 6. incidence, 7. sneezing, 8. congestion, 9. cough, 10. phlegm, 11. asthma, 12. conjunctivitis, 13. itchiness, 14. sinusitis, 15. bronchitis, 16. inflammation, 17. gastrointestinal, 18. diarrhea, 19. nausea, 20. stomachache, 21. dermatitis, 22. insomnia, 23. irritability, 24. appetite, 25. outbreak, 26. runny nose, 27. capture, 28. stimulate, 29. trigger, 30. exceed, 31. mumps, 32. measles, 33. bind, 34. receptor, 35. speculation, 36. inherit, 37. recessive, 38. dominant, 39. allele, 40. resistance, 41. gene, 42. pollution, 43. infection, 44. physiological, 45. autonomic, 46. genetic, 47. therapy, 48. pollinosis, 49.

ragweed, 50. mugwort, 51. inhale, 52. inhalation, 53. circulation, 54. meteorological, 55. prediction, 56. japanese cypress, 57. mediator, 58. inhibitor, 59. antagonist, 60. blocker, 61. steroid, 62. suppress, 63. histamine, 64. ameliorate, 65. desensitization, 66. treatment, 67. acclimatize, 68. consistent, 69. surgery, 70. nasal passage, 71. trend, 72. intertwine, 73. blank amazement, 74. male sterile, 75. propagate, 76. cell culture, 77. edible, 78. implementation, 79. staple food, 80. relieve, 81. severity, 82. toxicity, 83. carcinogenicity, 84. metastasis

Exercises 2

(1) F, (2) T, (3) F, (4) T, (5) F

Exercises 3

(1) The antigen-antibody reaction is a welcome function originally designed to protect humans.

(2) Some people develop bronchitis when nasal discharge enters the airway and causes inflammation.

(3) Environmental factors include exposure to the allergen, air pollution, infections, living environment, and eating habits.

(4) If the symptoms still do not subside, then we need to administer therapeutic drugs for hay fever.

(5) This product utilizes a desensitization method.

Exercises 4

(1) Refer to the original text.

(2) Refer to the original text.

(3) Not yet, there are complex processes on safety and efficacy assessments before commercialization.

(4) It is the metastasis of cancers.

(5) Discuss in your class using the text.

Chapter 7

Exercises 1

1. Fossil, 2. fossilize, 3. petroleum, 4. coal, 5. global warming, 6. mankind, 7. bless, 8. burn, 9. revolution, 10. population, 11. explode, 12. scorch, 13. carbon dioxide, 14. accompany by, 15. garner, 16. seaweed, 17. sewage, sludge, 18. fluctuation, 19. automobile, 20. foreign currency, 21. stalk, 22.

pomace, 23. crush, 24. molasse, 25. remain, 26. squeeze, 27. ferment, 28. output, 29. input, 30. contribution, 31. prevention, 32. reduction, 33. consume, 34. petrochemical, 35. synthetic, 36. commercialization, 37. monomer, 38. polymer, 39. polymerize, 40. biodegradable, 41. decompose, 42. microorganism, 43. disappear, 44. strength, 45. plasticity, 46. confer, 47. dispose of, 48. convert, 49. renewal, 50. sustainable, 51. principle, 52. collapse, 53. ecosystem, 54. emergence, 55. proliferate, 56. acceleration, 57. crucial, 58. deforestation, 59. reap

Exercises 2

(1) F, (2) F, (3) F, (4) F, (5) T

Exercises 3

(1) Biomass is translated in accord with its etymology of "living organism" (Bio) added to "quantity" (mass).

(2) Biofuels do not include fossil fuels such as petroleum, coal, and natural gas, which were originally plants that became fossilized.

(3) The rate of decomposition of PLA is said to be around 6 months to around 3 years.

(4) Bioethanol from sugarcane is the most popularly produced biofuel.

(5) The problem with biodegradable plastics is that their production requires a large amount of energy.

Exercises 4

(1) Global warming will be accelerated and we will worry on energy shortage as the fossil fuels are not originated from infinite sources.

(2) Refer to the text and think about saving, reducing and recycling on whatever you can.

(3) Because it is renewable and sustainable.

(4) Consider to make the energy use efficiency leading to higher degree of greenhouse gas reduction.

(5) We need many consideration : new policies, new systems, actions, education, social investment, infrastructure, and science & Technology with global consensus on governance.

Chapter 8

Exercises 1

1. uninvited, 2. jellyfish, 3. recollect, 4. sting, 5. bioluminescent, 6. limelight, 7. emit, 8. indispensable, 9. visible, 10. capture, 11. mature, 12. immense, 13. sensation, 14. lay eggs, 15. fertilize, 16. larva (e), 17. planulae, 18. spiral, 19. polyp, 20. seawall, 21. cilia, 22. nematocysts, 23. venom, 24. medusoid, 25. autonomous, 26. estuary, 27. subsequent, 28. embankment, 29. pier, 30. revetment, 31. massive, 32. eutrophication, 33. red tide, 34. zooplankton, 35. current, 36. reproductive, 37. miniscule, 38. astounding, 39. spawn, 40. proliferation, 41. contamination, 42. intimately, 43. coastline, 44. hospitable, 45. plug, 46. haul, 47. immeasurable, 48. invade, 49. mortality, 50. wither, 51. eelgrass, 52. prohibition, 53. devise, 54. ingenious, 55. contemplate, 56. shred, 57. trial-and-error, 58. prey on, 59. bountiful, 60. bumper crop, 61. profit, 62. pilchard (sardine), 63. predation, 64. suck in, 65. friction, 66. malfunction, 67. blink, 68. aggregate, 69. bottleneck, 70. collagen, 71. dietary, supplement, 72. purify, 73. lectin, 74. glycoprotein, 75. antibacterial, 76. mucin, 77. moisturize, 78. fertilizer, 79. pulsation, 80. arthritis, 81. plague, 82. stem

Exercises 2

(1) T, (2) F, (3) F, (4) T, (5) T

Exercises 3

(1) Echizen kurage achieves immense proportions when mature.
(2) This alone cannot explain the massive outbreaks of large jellyfish.
(3) A single spawning consists of at least 3 million eggs.
(4) Methods to prevent overfishing must also be considered.
(5) The nearly calorie-free jellyfish have also been dried and can be used as an ingredient in cookies.

Exercises 4

(1) Go to the text explanation.
(2) The reason is deeply related to the ocean environment.
(3) Go to the text and discuss.
(4) Go to the text and discuss.
(5) Firstly is to eat them. There are different uses : processing for functional food-additive substances, medical applications, cosmetics and

agroindustry.

Chapter 9

Exercises 1

1. bear, 2. nutritional, 3. hemisphere, 4. wind-pollinated, 5. insect-pollinated, 6. stamen, 7. pistil, 8. cross-fertilization, 9. European honeybee, 10. pollinator, 11. beekeeper, 12. pollinate, 13. antioxidant, 14. phenomenon, 15. spread, 16. radiation, 17. built-in, 18. characteristics, 19. alien, 20. abduction, 21. gorge, 22. hatch, 23. copulate, 24. hive, 25. acaricide, 26. eradicate, 27. arthropod, 28. regain, 29. expel, 30. acquire, 31. cat-and-mouse, 32. insecticide, 33. resistance, 34. organophosphate, 35. culprit, 36. nucleotide sequence, 37. decipher, 38. colony, 39. liquid nitrogen, 40. analyze, 41. conclude, 42. probability, 43. suspicion, 44. halt, 45. tremble, 46. paralysis, 47. microsporidian, 48. decisive, 49. reveal, 50. concomitant, 51. orchard, 52. countless, 53. self-incompatible, 54. pesticide, 55. stressful, 56. withstand, 57. duress

Exercises 2

(1) F, (2) T, (3) T, (4) T, (5) T

Exercises 3

(1) The search for the culprit was covered by the mainstream media on a near-daily basis.

(2) Despite the global popularity of almonds, their pollinators, the European honeybees, have started to disappear without warning.

(3) This mite attaches to a honeybee and sucks its body fluid.

(4) The entire nucleotide sequences of the European honeybee have been deciphered recently.

(5) Suspicion was settled on the Israel acute paralysis virus.

Exercises 4

(1) Fruit prices such as tomatoes, are increasing as a result.

(2) Possibilities would be given as : 1) "cell phone hypothesis" 2) "genetically engineered crop hypothesis", 3) "alien abduction hypothesis", 4) mite culprit hypothesis", 5) "Israel acute paralysis virus", 6) micro-sporidian, *Nosema apis* and 7) stress on intensive jobs at monoculture orchards.

(3) While the production can be seen in Greece, Spain Turkey and Australia, USA is the biggest producer.
(4) Multidisciplinary approaches are needed but yet no absolute solution has been provided so far as no one yet understands the underlying cause.
(5) It sounds like a scientific fiction.

Chapter 10

Exercises 1

1. ark, 2. endangered, 3. extinction, 4. implement, 5. ironically, 6. teeter, 7. brink, 8. paddy, 9. widespread, 10. drastic, 11. decline, 12. captivity, 13. enthusiasm, 14. wane, 15. reintroduce, 16. stuffed, 17. specimens, 18. preserve, 19. carcass, 20. conservationist, 21. foraging, 22. predator, 23. sociality, 24. fledgling, 25. acclimatize, 26. tragedy, 27. marten, 28. infiltrate, 29. immeasurable, 30. carnivorous, 31. crane, 32. nest, 33. omnivorous, 34. prey, 35. marsh, 36. deforestation, 37. autopsy, 38. corpse, 39. mercurial, 40. infertile, 41. abortive, 42. premature, 43. hatchling, 44. dragonfly, 45. eclosion, 46. eliminate, 47. pregnancy, 48. decoy, 49. inhabit, 50. equator, 51. wingspan, 52. abdominal, 53. ban, 54. steamboat, 55. vegetation, 56. exert, 57. trample, 58. indigenous, 59. landslide, 60. translocate, 61. vocalization, 62. volcanic, 63. earthquake, 64. eruption, 65. sanitation, 66. indigestion, 67. feral, 68. sanctuary, 69. cull, 70. crisis, 71. exalt

Exercises 2

(1) F, (2) T, (3) F, (4) F, (5) T

Exercises 3

(1) Although attempts to reduce the extinction rate by 2020 were implemented under Convention on Biological Diversity, they hardly had any impact.
(2) It is estimated that approximately 30 % of all bird species will die out by the end of the 21st century.
(3) One reason for this decline of individuals of Japanese crested ibis is thought to be that the crested ibises ate food that was contaminated by agricultural chemicals.
(4) Artificial breeding programs have been initiated on 1956.
(5) In the West, the stork is a lucky creature that delivers babies.

Exercises 4

(1) Human activities influenced strongly to the extinction.

(2) Those birds need training on foraging, flight, predator avoidance, sociality, and breeding.

(3) It is an exact model of a bird made of wood.

(4) Hunting was generally banned. For this reason, domestic wild bird species were protected.

(5) Discuss in your class.

Chapter 11

Exercises 1

1. deception, 2. wondrous, 3. mimicry, 4. walking stick, 5. branch, 6. emulate, 7. camouflage, 8. grasshopper, 9. surround, 10. predator, 11. prey, 12. assimilation, 13. tactic, 14. praying mantis, 15. twig, 16. congregate, 17. deceive, 18. hesitate, 19. bug, 20. monger, 21. excel, 22. juvenile, 23. molt, 24. chrysalis, 25. metamorphosis, 26. holometabolism, 27. instar, 28. pigment, 29. cryptic, 30. reveal, 31. gene expression, 32. unison, 33. eyespot, 34. melanism, 35. pollute, 36. lichen, 37. predominate, 38. verify, 39. phenomenon (na), 40. amid, 41. deceptive, 42. stinkbug, 43. undifferentiated, 44. prepupal, 45. ecdysteroid, 46. pupal, 47. apoptosis, 48. scale, 49. fragment, 50. secret, 51. solidify, 52. spin, 53. globule, 54. viscous, 55. scaffolding, 56. lure, 57. pheromone, 58. putrid, 59. stink, 60. olfaction, 61. emit, 62. obstacle, 63. hazard

Exercises 2

(1) T, (2) T, (3) T, (4) T, (5) F

Exercises 3

(1) The survival rate for black moths in the industrial area was higher than that of white moths.

(2) We have examined mimicry for avoiding detection by enemies with camouflage.

(3) After a few days, the moths were collected with a light trap and their numbers were examined.

(4) Butterflies are known as mimics not only as adults but from the larval stage as well.

(5) *Drakea* orchid mimics a female wasp and makes the congregating male wasps serve as pollinators.

Exercises 4

(1) Mimicry can be divided into 1) camouflage, 2) standard mimicry, 3) aggressive mimicry, and 4) reproductive mimicry. Mimicry can also be classified into 1) Batesian mimicry and 2) Mullerian mimicry.
(2) It is a built-in talent for survival.
(3) Discuss the mechanisms in your class.
(4) Industrial revolution has been taken place and pollution with soot because coal was used in massive quantities as a power source.
(5) Consult with the text.

Chapter 12

Exercises 1

1. genome, 2. model organism, 3. fruit fly, 4. rhesus macaque, 5. primate, 6. amphibian, 7. African clawed frog, 8. zebrafish, 9. sea squirt, 10. backdrop, 11. brook, 12. humorous, 13. countenance, 14. brackish water, 15. hibernate, 16. lifespan, 17. omnivorous, 18. zooplankton, 19. phytoplankton, 20. tail fin, 21. dorsal fin, 22. pectoral fin, 23. pelvic fin, 24. spawn, 25. adhere, 26. abdomen, 27. layperson, 28. chromosome, 29. fertilize, 30. ovum, 31. sexual differentiation, 32. germ cell, 33. testis (testes), 34. genital primordium, 35. embryo, 36. ovary, 37. estrogen, 38. stroid, 39. gonadal somatic cell, 40. testosterone, 41. microscope, 42. ova, 43. sperm, 44. masculinize, 45. endocrine, 46. overturn, 47. perceive, 48. trichromatic, 49. mammal, 50. dichromatic, 51. treetop, 52. creature, 53. turbidity, 54. fluctuation, 55. variegate, 56. aqueous, 57. retrogression, 58. implication, 59. lineage, 60. genetic duplication, 61. clue, 62. vertebrate, 63. decipher, 64. silkworm, 65. cascade, 66. longitudinal

Exercises 2

(1) T, (2) T, (3) F, (4) F, (5) F

Exercises 3

(1) Medaka's lifespan is approximately 1 to 3 years.
(2) Before the Medaka genome deciphered, zebrafish as a model organism, had been at the center of genomic research.

(3) Sexual differentiation of the medaka is what first attracted attention.

(4) A comparison of the human and medaka genomes shows that 60 % of the genes are resembled.

(5) The vertebrates include fish, amphibians, reptiles, birds, and mammals.

Exercises 4

(1) Examine the text.

(2) Look the biological features of the species for easy experimental handling.

(3) Yes, read the text.

(4) Discuss in your class.

(5) It can contribute to evolutionary studies of vertebrates and many applications to comparative biology including medical applications.

Index

—— 編著者・著者略歴 ——

渡邉　和男（わたなべ　かずお）

1983 年　神戸大学農学部園芸農学科卒業
1985 年　神戸大学大学院修士課程修了
1988 年　ウィスコンシン大学大学院博士課程修了（遺伝育種学専攻）
　　　　Ph. D.（ウィスコンシン大学）
1988 年　国際ポテトセンター主任研究員
1991 年　コーネル大学助教授
1996 年　近畿大学助教授，国際植物遺伝資源研究所（IPGRI）名誉研究員
2001 年　筑波大学教授，コーネル大学在外特別教授
2004 年　筑波大学大学院教授
　　　　現在に至る

渡邉　純子（わたなべ　じゅんこ）

1986 年　神戸大学教育学部初等教育学科卒業
1986 年　兵庫県立須磨東高等学校教諭
1995 年　コーネル大学大学院修士課程修了（園芸，植物生理学専攻）
1997 年　近畿大学非常勤講師
2001 年　筑波大学遺伝子実験センター研究推進員
2005 年　農業生物資源研究所非常勤職員
2007 年　科学作家
　　　　現在に至る

続 英語で学ぶ生物学
—— 生物科学の新しい挑戦 ——
The Second Volume of Endeavors in Biological Sciences
—— New Challenges in Biological Sciences ——

© Kazuo Watanabe, Junko Watanabe 2013

2013 年 11 月 8 日　初版第 1 刷発行　★

検印省略

編著者	渡　邉　和　男	
著　者	渡　邉　純　子	
発行者	株式会社　コロナ社	
	代表者　牛来真也	
印刷所	萩原印刷株式会社	

112-0011　東京都文京区千石 4-46-10
発行所　株式会社　**コ　ロ　ナ　社**
CORONA PUBLISHING CO., LTD.
Tokyo　Japan
振替 00140-8-14844・電話(03)3941-3131(代)
ホームページ　http://www.coronasha.co.jp

ISBN 978-4-339-07795-7　（高橋）　（製本：愛千製本所）
Printed in Japan

本書のコピー，スキャン，デジタル化等の無断複製・転載は著作権法上での例外を除き禁じられております。購入者以外の第三者による本書の電子データ化及び電子書籍化は，いかなる場合も認めておりません。

落丁・乱丁本はお取替えいたします